Alexandra Beate Suchomel

Food availability in Austria based on household budget surveys

Alexandra Beate Suchomel

Food availability in Austria based on household budget surveys

Südwestdeutscher Verlag für Hochschulschriften

Impressum / Imprint
Bibliografische Information der Deutschen Nationalbibliothek: Die Deutsche Nationalbibliothek verzeichnet diese Publikation in der Deutschen Nationalbibliografie; detaillierte bibliografische Daten sind im Internet über http://dnb.d-nb.de abrufbar.
Alle in diesem Buch genannten Marken und Produktnamen unterliegen warenzeichen-, marken- oder patentrechtlichem Schutz bzw. sind Warenzeichen oder eingetragene Warenzeichen der jeweiligen Inhaber. Die Wiedergabe von Marken, Produktnamen, Gebrauchsnamen, Handelsnamen, Warenbezeichnungen u.s.w. in diesem Werk berechtigt auch ohne besondere Kennzeichnung nicht zu der Annahme, dass solche Namen im Sinne der Warenzeichen- und Markenschutzgesetzgebung als frei zu betrachten wären und daher von jedermann benutzt werden dürften.

Bibliographic information published by the Deutsche Nationalbibliothek: The Deutsche Nationalbibliothek lists this publication in the Deutsche Nationalbibliografie; detailed bibliographic data are available in the Internet at http://dnb.d-nb.de.
Any brand names and product names mentioned in this book are subject to trademark, brand or patent protection and are trademarks or registered trademarks of their respective holders. The use of brand names, product names, common names, trade names, product descriptions etc. even without a particular marking in this work is in no way to be construed to mean that such names may be regarded as unrestricted in respect of trademark and brand protection legislation and could thus be used by anyone.

Coverbild / Cover image: www.ingimage.com

Verlag / Publisher:
Südwestdeutscher Verlag für Hochschulschriften
ist ein Imprint der / is a trademark of
OmniScriptum GmbH & Co. KG
Heinrich-Böcking-Str. 6-8, 66121 Saarbrücken, Deutschland / Germany
Email: info@svh-verlag.de

Herstellung: siehe letzte Seite /
Printed at: see last page
ISBN: 978-3-8381-5175-5

Zugl. / Approved by: Wien, Universität Wien, Diss., 2008

Copyright © 2015 OmniScriptum GmbH & Co. KG
Alle Rechte vorbehalten / All rights reserved. Saarbrücken 2015

MY SPECIAL THANKS

Prof. Dr. Ibrahim ELMADFA

My special thanks go to my Supervisor Prof. Dr. Ibrahim ELMADFA who made this doctoral thesis possible. The Austrian participation in the project "Data Food Networking" would not be possible and this dissertation would not be accomplished without his support. Sincere thanks, that I was allowed to take an active part in the DAFNE IV – meetings and that I was allowed to represent the Institute of Nutritional Sciences of the University of Vienna.

DAFNE – COORDINATING CENTRE

Best thanks to the DAFNE – Coordinating Centre in Athens who accompanied me with advices and critical explanatory notes within the project work.

TABLE OF CONTENTS

LIST OF FIGURES	VI
LIST OF TABLES	XI
ABBREVIATIONS	XIII

1	INTRODUCTION	1
2	LITERATURE REVIEW AND OBJECTIVES	4
2.1.	BACKGROUND – The DAFNE Project	5
2.1.1.	Individual dietary surveys	7
2.1.2.	Household Budget Surveys	8
2.1.3.	Food Balance Sheets	10
2.2.	AVAILABILITY	12
3	DAFNE – DAta Food NEtworking	14
3.1.	HBS HISTORY BACKGROUND	14
3.2.	Austrian Household Budget Survey - Konsumerhebung	17
3.3.	DAFNE history	19
3.3.1.	DAFNE I	20
3.3.2.	DAFNE II	22
3.3.3.	DAFNE III	23
3.3.4.	DAFNE IV	24
3.3.5.	DAFNE V	24
3.3.6.	DAFNE databank	25
3.4.	COICOP – Classification of Individual Consumption of Purpose	27
4	MATERIAL, dataset used	29
4.1.	STATISTICAL METHODS	39
4.1.1.	Conversion of recorded expenses into quantities on	

TABLE OF CONTENTS

	the basis of national annual production quantities	39
4.1.2.	Conversion of expenses by the price per unit weight to replace zero values for quantities	40
5	METHODOLOGY	43
5.1.	Austrian HBS data collection methodology	43
5.2.	DAFNE TASKS	46
5.3.	Creation of comparable tables regarding socio-demographic parameters and food groups available in the HBS datasets	51
5.3.1.	Definition of comparable categories of socio-demographic data, among the DAFNE-member countries	51
5.3.1.1.	Category of locality	53
5.3.1.2.	Education of household head	54
5.3.1.3.	Occupation of household head	56
5.3.1.4.	Category of household composition	60
5.3.2.	Definition of comparable categories of food items among North-, Central- and South European countries	61
6	VALIDATION	80
6.1.	Comparison between Austrian results of DAFNE-Household Budget Survey, Austrian Study on Nutritional Status and FBS	80
6.1.1.	FBS	80
6.1.2.	Austrian Study on Nutritional Status (ASNS)	81
6.2.	Conclusion for the validation	83
7	RESULTS – FOOD AVAILABILITY IN AUSTRIA	86
7.1.	Mean Availability in Austria	86

TABLE OF CONTENTS

7.2.	Mean Availability by educational attainment in Austria	93
7.3.	Mean Availability by locality in Austria	101
7.4.	Mean Availability by occupation	107
7.5.	Mean Availability by household composition	118
8	FOOD AVAILABILTY IN EUROPE	132
8.1.	Mean daily food availability among elderly European living alone	136
8.2.	Cultural and Social factors	137
8.3.	European Food Availability	138
8.3.1.	Food Availability in North, Central and South Europe - Finland, Austria, Greece	140
8.3.2.	DAFNE members in comparison to mean DAFNE food availability	169
8.3.3.	DAFNE members in comparison to Austrian mean food availability	176
9	DISCUSSION	200
9.1.	HBS versus INS	200
9.2.	HBS versus FBS	200
9.3.	HBS versus FBS and INS	201
9.4.	Recording period	204
9.5.	The influence of socio-economic parameters	205
9.6.	LIMITATIONS	214
9.6.1.	Underreporting	214
9.6.2.	The need for a unified HBS food coding system	216
9.6.3.	Purchases – food availability	218
9.6.4.	Larder stocks	218
9.6.5.	Waste and visitors´ consumption of food	219

TABLE OF CONTENTS

9.7.	NUTRIENTS	219
9.8.	FOOD EATEN OUTSIDE THE HOME	222
9.8.1.	DAFNE IV and meals out of home	225
9.9.	AVERAGE DAILY AVAILABILITY IN THE DAFNE COUNTRIES, IN THE 1990s	227
10	CONCLUSIONS AND RECOMMENDATIONS	238
10.1.	Austrian trends in daily food availability	238
10.2.	Association of AUSTRIAN food availability with socioeconomic characteristics	238
10.3.	Disparities in food patterns in Europe	240
10.4.	Austrian results in comparison to those of the other DAFNE-countries in the nineties	242
10.5.	Information of food availability collected by HBS, advantages and disadvantages	243
11	SUMMARY / ZUSAMMENFASSUNG	247
12	REFERENCES	251

TABLE OF CONTENTS
ANNEXES

ANNEX 1	Graphs of food and beverage availability by DAFNE food groups in Europe	A1
ANNEX 2	Detailed tables of results	A13
ANNEX 2.1.	Tables of food availability by locality	A13
ANNEX 2.2.	Tables of food availability by education	A17
ANNEX 2.3.	Tables of food availability by occupation	A21
ANNEX 2.4.	Tables of food availability by household composition	A26
ANNEX 3	Graphs on food availability by socio-demographic characteristics	A30
ANNEX 4	Available information of occupation in the Austrian HBS	A59
ANNEX 5	Classification of Individual Consumption by Purpose	A63

List of Figures

Figure 1.	Timing of the national surveys	14
Figure 2.	Collaboration framework for HBS	16
Figure 3a.	Housekeeping book	33
Figure 3b.	Housekeeping book	34-35
Figure 4.	DAFNE methodology	50
Figure 5.	Definition of urbanisation	53
Figure 6.	Definition of education	55
Figure 7.	Definition of occupation	57-59
Figure 8.	Definition of composition of household	60
Figure 9.	Total food expenditure by age of housewife	85
Figure 10.	Mean availability of food groups of plant origin in Austria	87
Figure 11.	Mean availability of meat and meat products	88
Figure 12.	Mean availability of potatoes and vegetables	89
Figure 13.	Mean availability of fruits	90
Figure 14.	Mean availability of dairy products	91
Figure 15.	Mean availability of fats and oils	91
Figure 16.	Mean availability of beverages and stimulants	92
Figure 17.	Average food availability, by education of the household head	94
Figure 18.	Biscuits, cookies, cakes, rusks, by education	95
Figure 19.	Fresh and preserved milk, by education	96
Figure 20.	Yoghurt, ice-cream, other milk products, by education	97
Figure 21.	Fruit and vegetable juices, by education	100
Figure 22.	Average food availability, by locality	102
Figure 23.	Average food availability, by locality	102
Figure 24.	Average food availability, by locality	103
Figure 25.	Potatoes and tuber plants, by locality	104
Figure 26.	Fruit Lipids of animal and vegetables origin, by locality	104
Figure 27.	Fresh and preserved milk, by locality	105
Figure 28.	Average food availability, by occupation of the household head	109
Figure 29.	Cheese and curd cheese, by occupation	111
Figure 30.	Average food availability, by occupation of the household head	113
Figure 31.	Margarine, vegetable oils and olive oil, by occupation	115
Figure 32.	Fish and seafood, by occupation	116
Figure 33.	Average food availability, by household composition	121
Figure 34.	Average food availability, by household composition	122
Figure 35.	Average food availability, by household composition	123
Figure 36.	Red meat, poultry, meat products and offals, by household composition	124
Figure 37.	Chocolate, sweets, jam and honey, by household composition	125

LIST OF FIGURES

Figure 38.	Average food availability, by household composition	126
Figure 39.	Rice and other cereal products, by household composition	127
Figure 40.	Determinants of food consumption potential	133
Figure 41.	Mean daily individual availability of selected foods among elderly Europeans living alone	137
Figure 42.	DAFNE main food groups in North, Central and South Europe	141
Figure 43.	Daily availability of potatoes – mean	142
Figure 44.	Daily availability of potatoes, by education	142
Figure 45.	Daily availability of meat and meat products	144
Figure 46.	Daily availability of fruits – mean	145
Figure 47.	Daily availability of vegetables, by household composition	147
Figure 48.	Daily availability of vegetables, by occupation	147
Figure 49.	Daily availability of bread and rolls, by occupation	148
Figure 50.	Daily availability of non alcoholic beverages – mean	150
Figure 51a.	Distribution of alcoholic beverages in North, Central, South Europe	151
Figure 51b.	Distribution of alcoholic beverages in North, Central, South Europe	152
Figure 51c.	Distribution of alcoholic beverages in North, Central, South Europe	152
Figure 52.	Mean daily availability in North, Central, and South Europe	154
Figure 53.	Mean daily availability of milk products, by education	155
Figure 54.	Daily availability of milk products, by locality	156
Figure 55a.	Distribution of total lipids in North, Central, South Europe	157
Figure 55b.	Distribution of total lipids in North, Central, South Europe	158
Figure 55c.	Distribution of total lipids in North, Central, South Europe	158
Figure 56.	Mean daily availability	165
Figure 57.	Eggs – mean	170
Figure 58	Lipids of (a) animal origin and (b) vegetable origin	171
Figure 59.	Potatoes – mean	172
Figure 60.	(a) Cereals and cereal products and (b) Flour	173
Figure 61.	(a) Cheese and (b) Pasta	175
Figure 62a.	Sugar by Elementary education	176
Figure 62b.	Sugar by Secondary education	177
Figure 62c.	Sugar by Higher education	177
Figure 63a.	Cereals and Products by Elementary education	178
Figure 63b.	Cereals and Products by Secondary education	179
Figure 63c.	Cereals and Products by Higher education	180
Figure 64a.	Milk, by Manual household head	181
Figure 64b.	Milk, by Unemployed household head	182
Figure 65a.	Milk products in Rural areas	183
Figure 65b.	Milk products in Urban areas	184
Figure 66.	Lipids of vegetable origin – mean	185
Figure 67a.	Lipids of animal origin Rural areas	186

LIST OF FIGURES

Figure 67b.	Lipids of animal origin Urban areas	187
Figure 68a.	Lipids of animal origin "adult-single" HH	188
Figure 68b.	Lipids of animal origin "adult-2-members" HH	189
Figure 69a.	Lipids of animal origin "elderly-single" HH	190
Figure 69b.	Lipids of animal origin by "elderly-2 members" HH	191
Figure 70a.	Lipids of vegetable origin in Rural areas	192
Figure 70b.	Lipids of vegetable origin in Urban areas	193
Figure 71a.	Lipids of vegetable origin by "adult-single" HH	194
Figure 71b.	Lipids of vegetable origin by "adult-2 members" HH	195
Figure 72a.	Lipids of vegetable origin "elderly-single" HH	196
Figure 72b.	Lipids of vegetable origin by "elderly-2 members" HH	197
Figure 73a.	Fish and Seafood by Retired household head	198
Figure 73b.	Fish and Seafood by Unemployed household head	199
Figure 74.	Mean Daily Per Capita Availability of Total Added Lipids and Olive Oil in Austria, Finland, Greece, Spain	208
Figure 75a.	Life expectancy at birth of men in North-, Central, South-Europe	210
Figure 75b.	Life expectancy at birth of women in North-, Central, South-Europe	211
Figure 76a.	Life expectancy at age 60 of men in North-, Central, South-Europe	212
Figure 76b.	Life expectancy at age 60 of women in North-, Central, South-Europe	213
Figure 77.	Food consumption expenditures of households in Austria	225
Figure 78.	Daily availability of fruits, by occupation	A30
Figure 79.	Daily availability of fruits, by education	A30
Figure 80.	Daily availability of vegetables, by locality	A31
Figure 81.	Daily availability of vegetable oils, by mean	A31
Figure 82.	Daily availability of vegetable fats, by education	A32
Figure 83.	Daily availability of lipids of animal origin, by education	A32
Figure 84.	Daily availability of non alcoholic beverages, by education	A33
Figure 85.	Daily availability of non alcoholic beverages, by occupation	A33
Figure 86.	Daily availability of non alcoholic beverages, by household composition	A34
Figure 87.	Daily availability of alcoholic beverages, by occupation	A35
Figure 88.	Daily availability of alcoholic beverages, by education	A35
Figure 89.	Daily availability of wine, by occupation	A36
Figure 90.	Daily availability of beer, by occupation	A36
Figure 91.	Daily availability of spirits, by occupation	A37
Figure 92.	Daily availability of bread and rolls, by household composition	A37
Figure 93a.	Potatoes, by "adult-single"	A38
Figure 93b.	Potatoes, by "elderly-single"	A38
Figure 94a.	Potatoes, by "elderly-2 members"	A39

LIST OF FIGURES

Figure 94b.	Potatoes, by "adult-2 members"	A39
Figure 95a.	Flour, manual household head	A40
Figure 95b.	Flour, non manual household head	A40
Figure 96a.	Coffee in Rural areas	A41
Figure 96b.	Coffee in Urban areas	A41
Figure 97a.	Spirit in Rural areas	A42
Figure 97b.	Spirit in Urban areas	A42
Figure 98a.	Spirit, by "adult-single"	A43
Figure 98b.	Spirit, by "adult-2 members"	A43
Figure 99a.	Spirit, by "elderly-single" households	A44
Figure 99b.	Spirit, by elderly-2 members households	A44
Figure 100a.	Beer in Rural areas	A45
Figure 100b.	Beer in Rural Urban areas	A45
Figure 101a.	Beer, by "adult-single" households	A46
Figure 101b.	Beer, by "adult-2 members" households	A46
Figure 102a.	Beer, by "elderly-single" households	A47
Figure 102b.	Beer, by "elderly-2 members"	A47
Figure 103a.	Wine in Rural areas	A48
Figure 103b.	Wine in Urban areas	A48
Figure 104a.	Wine, by "adult-single" households	A49
Figure 104b.	Wine, by "adult-2 members" households	A49
Figure 105a.	Wine, by "elderly-single" households	A50
Figure 105b.	Wine, by "elderly-2 members" households	A50
Figure 106a.	Milk, by "adult-single" households	A51
Figure 106b.	Milk, by "adult-2 members" households	A51
Figure 107a.	Milk, by "elderly-single" households	A52
Figure 107b.	Milk, by "elderly-2 members" households	A52
Figure 108a.	Bread and Rolls, by "adult-single" households	A53
Figure 108b.	Bread and Rolls, by "adult-2 members" households	A53
Figure 109a.	Bread and Rolls, by "elderly-single" households	A54
Figure 109b.	Bread and Rolls, by "elderly-2 members" households	A54
Figure 110a.	Cocoa in Rural areas	A55
Figure 110b.	Cocoa in Urban areas	A55
Figure 111a.	Eggs, by Elementary education	A56
Figure 111b.	Eggs, by Secondary education	A56
Figure 111c.	Eggs, by Higher education	A57
Figure 112a.	Mineral water, by Elementary education	A57
Figure 112b.	Mineral water, by Secondary education	A58
Figure 112c.	Mineral water, by Higher education	A58
Figure 113.a.	Available information of occupation in the Austrian HBS – Code I, Code II	A59
Figure 113.b.	Available information of occupation in the Austrian HBS – Code II, Code III, Code IV	A60

LIST OF FIGURES

Figure 113.c.	Available information of occupation in the Austrian HBS – Code V, Position in Occupation 0-15	A60
Figure 113.d.	Available information of occupation in the Austrian HBS – Code V, Position in Occupation 0-15	A61
Figure 113.e.	Available information of occupation in the Austrian HBS – Job description	A62
Figure 114.	COICOP – HBS 03	A63-A70

List of Tables

Table 1.	Distribution of households participating in the 1999/2000 Austrian HBS, by federal states	36
Table 2.	Reference Period and the contributive households	38
Table 3.	Minced Meat	62
Table 4.	Comparable groups of cereals and cereal products	63-64
Table 5.	Comparable groups of meat, meat products and dishes	64-66
Table 6.	Comparable groups of fish, seafood and dishes	66-67
Table 7.	Comparable groups of eggs, milk and milk products	67-68
Table 8.	Comparable groups of added lipids	69-70
Table 9.	Comparable groups of potatoes and other starchy roots, pulses and nuts	71
Table 10.	Comparable groups of vegetables	72-73
Table 11.	Comparable groups of fruits	73-74
Table 12.	Comparable groups of fruit and vegetable juices	75
Table 13.	Comparable groups of sugar and sugar products	75
Table 14.	Comparable groups of non-alcoholic beverages	76
Table 15.	Comparable groups of alcoholic beverages	77
Table 16.	Comparable groups of miscellaneous and dishes	78-79
Table 17	Comparison between DAFNE, ASNS and FBS results for Austria	82
Table 18.	DAFNE main food groups in Austria	86
Table 19.	DAFNE main food groups, by education in Austria	93
Table 20.	DAFNE main food groups, by locality in Austria	101
Table 21.	DAFNE main food groups, by occupation in Austria	107
Table 22a.	DAFNE main food groups, by household composition in Austria	118
Table 22b.	DAFNE main food groups, by household composition in Austria	119
Table 23.	Mean daily food availability in Finland, Austria, Greece	168
Table 24.	Average daily availability of fresh fruits by type in the DAFNE countries, in the 1990	228
Table 25.	Average availability of total meat and meat products in the 1990s	229
Table 26.	Average daily availability of vegetables in the DAFNE countries, in the 1990s	230
Table 27.	Daily Availability of cheese in the DAFNE countries in the 1990s	231
Table 28.	Daily availability of milk	232

LIST OF TABLES

Table 29.	Daily availability of cheese and milk products	233
Table 30.	Daily availability of sugar and sugar products	234
Table 31.	Daily availability of potatoes	235
Table 32.	Daily availability of nuts	236
Table 33.	Daily availability of pulses by mean	237

LIST OF ABBREVIATIONS

A	Austria
B	Belgium
D	Germany
DAFNE	Data Food Networking
E	Spain
ENHR	European Nutrition and Health Report
EU	European Union
FAO	Food and Agriculture Organization
FBS	Food Balance Sheets
Fig.	Figure
FIN	Finland
g/p/d	gram per person per day
GR	Greece
H	Hungary
HBS	Household Budget Survey(s)
HH	Households
I	Italy
INN-CA	Italian National Food Survey
INS	Individual Nutrition Surveys
IRL	Ireland
L	Luxembourg
ml/p/d	millilitre per person per day
N	Norway
p.	page
P	Portugal
PL	Poland
S	Sweden
UK	United Kingdom

1 INTRODUCTION

Diets and their nutritional value changed dramatically during the 20th Century as a result of agricultural and industrial progress, commercialisation of the food supply and changing lifestyles for different groups of the population.

Food consumption and nutrient intake vary across Europe, within countries and regions, given the wide range of cultural, psycho-social, and practical factors which can influence food behaviour. One of the cultural richness of the European Union is its culinary variety with very different food behaviour in Europe, within countries regional and socio-economic variations.

Published data on health inequalities have shown that people who are poorer, have lower educational levels and lower income, are also disadvantaged in health and life expectancy.

Variations in health inequalities among different countries have often been explained by differences in welfare policies and living standards. Reasons for health inequalities are complex and can be affected by economic, cultural and personal factors. The level of inequality in material resources within a society has often been presented as a major cause of health inequality. Living and working conditions of those belonging to certain social groups expose them to greater health hazards, because of cultural, behavioural and psychosocial factors. Europeans belonging to higher social class groups tend to have healthier diets, for instance consume more fruits and vegetables, which are important sources of antioxidants and micronutrients, are central in the prevention of non-communicable

INTRODUCTION

diseases, such as cardiovascular diseases and cancer. In Europe in the North and West, those people with a high educational level tend to consume more vegetables and fruits, i.e. have healthier food habits than those with a low educational level. On the other hand in southern European countries elementary education of the household head was found to be associated with higher vegetable availability.

An increase in the availability of fruit, and vegetables in many Northern and Western countries has been documented in recent years. In Southern Europe, fruit and vegetable intakes have not shown the same increase, although they began at a much higher level (Elmadfa I. et al., 2005a, p.15). The role of fruits and vegetables is probably more traditional in the South. In regions where consumption of vegetables and fruits is more common, the lower social classes tend to consume more of these than higher social classes. Regional variations are linked to structural characteristics such as availability of fresh vegetables. Those of lower socio-economic classes in the South may have better access to cheaper vegetables and fruits. (Roos G. et al., 2001, p.36-40).

Dietary patterns and nutritional intakes vary across Europe and within countries. This variability reflects the multitude of cultural, economic, geographic and psycho-social factors that can influence food behaviour.

Comparable between countries information on food availability can be provided by data collected in the household budget surveys (HBS). The HBS collect data on the quantities and values of all food items available, among nationally representative samples of households. No allowance is made for food wasted or given to pets. Individual values were also estimated, without taking into account age and sex differences of the

INTRODUCTION

household members. DAFNE is the acronym for DAta Food NEtworking and aims at the creation of a pan-European food data bank based on HBS. The DAFNE databank may serve as a tool for identifying, quantifying and depicting variation of food habits in Europe (Trichopoulou A., 2001, p.1187).

Despite the increasing importance of nutrition in integrated policy initiative to tackle the rise in non-communicable diseases worldwide, a large amount of the existing evidence for the role of dietary intake in explaining worldwide differences in mortality and disease incidence is based on ecological information, with food availability being used as a proxy for dietary intake.

Differences in dietary patterns among regions and countries are acknowledged, assessing the level of these differences represents a major challenge for researchers.

While it is generally acknowledged that Food Balance Sheets (FBS) tend to overestimate dietary intakes in developed countries and possibly underestimate intakes in less developed countries, little information is available on the magnitude of the discrepancy between food supply data and information from dietary surveys of individual intakes with regards to various types of foods (Pomerleau J. et al., 2003, p.827). In general, food supply is clearly above actual intake availability. This fact was shown in the ENHR 2004 by comparing fruit and vegetable supply (353 and 262 g/p/d, respectively), availability (192 and 142 g/p/d, respectively) and intake (183 and 148 g/p/d, respectively) in Austria (Elmadfa I. et al., 2005a, p.63).

2 LITERATURE REVIEW AND OBJECTIVES

This thesis demonstrates the use of HBS for representing the Austrian food availability: the average daily availability, availability by socio-demographic parameters and the Austrian food data in comparison to DAFNE members.

Further the HBS data and the harmonization of selected socio-demographic and food data of three European countries will be used for representing the methodological approach: Finland, as a representative of the North European countries, Austria, as a typical Central European country, and Greece with the characteristics of a typical Mediterranean country.

The results will be presented and the limitations and perspectives will be discussed.

Since the Austrian results and deliverables of DAFNE IV project are enormously extensive and huge, an excerpt of Austrian food availability is presented in chapter 6 (socio-demographic comparisons of food consumption within Austria) and in chapter 7 (comparisons of Austrian food consumption with DAFNE countries).

Further illustrations and graphs of Austrian food consumption based on HBS are presented in ANNEX 1 food availability in DAFNE Europe) and in ANNEX 3 (food availability by socio-demographic characteristics).

Detailed and comparative HBS values of Austrian food availability are listed in tables in the respective chapter of socio-demographic parameters of Austria (chapter 6.1. – 6.5.), in a table which presents a comparison of the Central European DAFNE country Austria with North and South DAFNE Europe (chapter 7.3.1.) and in tables of DAFNE food subgroups by socio-demographic characteristics (ANNEX 2).

2.1. BACKGROUND – THE DAFNE PROJECT

There appears to be nutritional variation within countries by socio-demographic groups, defined by their residence and educational level. The distribution patterns of food availability provide insights into the socio-economic determinants of food preferences, as conditioned by market forces (Trichopoulou A., 2001, p.1187).

The DAFNE databank allows the assessment of trends in the food habits of European populations, as well as comparisons between population subgroups defined according to their demographic and socio-economic characteristics (Trichopoulou A. et al., 2005, p.69).

National nutrition surveillance programmes should rely on data collected in the context of nation-representative and regularly conducted individual nutrition surveys (INS).

These surveys are undertaken only in a limited number of countries, because there are expensive and labour intensive (Naska A. et al., 2001b, p.1159). The prohibitive cost of implementing special dietary surveys may limit the European coverage of data collection. In the modern world of rapid changes however, nutrition monitoring should make use of dietary surveys that have built-in mechanism of continuity over time, extensive coverage and allow international comparisons (Trichopoulou A. and Naska A., 2003a, p.27). These countries usually have robust economies and years of experience in the field of dietary monitoring. Furthermore variable dietary assessment methods are used, making difficult to accomplish comparability at the national level (Naska A., 2001b, p.1159), the representativeness of the survey population and the nature of the data (e.g. reflecting or not habitual intake) may also challenge the suitability of a dietary survey for international comparisons (Trichopoulou A. and Naska A., 2003a, p.27).

A national nutrition information system is essential for an effective national nutrition policy (Australian Institute, 1996, p.3).

The parameters (among a series of others available)"general information", "information on food", "sociodemographic information" and "information on household finances" from the HBS's data of Austria collected from the Austrian Statistical Office, which are for the DAFNE project requested, are listed in chapter 3.Material.

Due to different systems in regard of taxes, salaries among countries, it is difficult to compare objectively the "real" income and expenditure of the households. Complex estimations are necessary before achieving an insight, which would allow comparisons among countries (Kanellou A., 1999, p.7-8).

It is important to recognize the limitations of the information on food consumption because the level at which the data were obtained limits the types of dietary change which can be observed and the conclusions regarding the factors involved (Meiselman H.L., 1996, p.366).

Each type of dietary information data corresponds to a different stage in food chain and is obtained by different methods (Kroes R. et al., 2002, p.343).

HBS refer to the beginning of the dietary chain (purchased food items brought into the household), while INS refer to the end of it (food prepared, cooked and consumed) (Naska A. et al., 2001a, p.1155).

The methodology of food-data collection differs in information, detail and accuracy of random sampling. Food-data can be collected on three levels:

INS data present information about the effectively food consumption of individuals. Additional information can be collected about food consumption habits.

Food-data based on HBS deliver information about purchases in expenses and quantities, but the HBS do not predicate about the handling of food in the household and how far products are definitively consumed.

Food-data on national level represent information on the kind and the quantity of food, which are available for consumption. National surveys do not represent the distribution of food on several subgroups. The national food quantity, which is available for consumption, is calculated in FBS. These data are collected for the majority of countries in the world by the Food and Agriculture Organization of the United Nations (FAO) (Becker W. and Helsing E., 1993, p.22).

2.1.1. Individual dietary surveys

In contrast to FBS and HBS, intake data at the individual level provide information on average food and nutrient and their distribution over various well-defined groups of individuals. These data more closely reflect actual consumption. To collect dietary intake data at an individual level, several methods can be used. Briefly, the methods can be divided into two

LITERATURE REVIEW AND OBJECTIVES

categories: record and recall methods. Record methods collect information on current intake over one or more days. Recall methods reflect past consumption, varying from intake over the previous day (24-hour recall) to usual food intake (dietary history or food frequency) (Kroes R. et al., 2002, p.346).

Data on the individual level facilitate estimation of the adequacy of dietary intake and studying the relationship of diet and health. Therefore, for the monitoring of relevant dietary indicators in Europe, data at the individual level are preferred (Biro G. et al., 2002, p.26).

2.1.2. Household Budget Surveys

Comparable between-countries information on food availability can also be provided by data collected in HBS (Naska A. et al., 2001b, p. 1159).

In the HBS a record is kept on the quantity of food that entered the household together with the number of household members consuming the food (Kanellou A., 1999, p. 86).

The HBS, which are carried out occasionally, provide information on food consumption at the household level in the country. The HBS usually collect information from different sections of the population. As a result, they are able to provide indications of the diet obtained by people living in different parts of a country, or by people of different socio-economic groups. If the surveys are carried out through the year, they give information about seasonal variation, and if they are carried out at regular intervals they enable longer term trends to be detected and analyzed.

LITERATURE REVIEW AND OBJECTIVES

Household food consumption patterns are the product of diverse set of driving factors that shape the day-to-day and long-term decisions that consumers make. It is possible to outline and group these driving factors in a number of ways, probably none of which captures each and every factor nearly without overlap or omissions (OECD, 2001, p.18).

HBS collect data on food availability at household level (Trichopoulou A. and Naska A., 2003a, p.25), taking into consideration the purchases, contributions from their own production and food items offered to members as gifts during the reference period. HBS are systematically conducted by the National Statistical Offices (Trichopoulou A. et al., 2005, p.70).

The use of HBS data for nutritional purposes is highly cost-effective. This information, which is regularly collected and updated, allows a low cost assessment of trends in food habits and the identification of dietary habits of population subgroups (Rodrigues SSP. and de Almeida MDV., 2001, p.1167).

HBS can form the basis for a database that meets several criteria of a contemporary nutritional database that needs to be

1.) truly international
2.) representative and linked to explanatory demographic and socioeconomic factors that are themselves subject to rapid changes
3.) very large, to generate precise estimates for inherently complex patterns
4.) regularly updated
5.) affordable (Trichopoulou A. et al., 1999, p.129)

2.1.3. Food Balance Sheets

FBS are constructed by the FAO from national accounts of the supply and use of foods. Data are calculated from the food produced and imported for countries as a whole, subtracting food exported, fed to animals, or otherwise not available to humans, and dividing by the population to obtain average values per person/ year. In wealthy countries, food availability estimated by FBS are clearly overestimated, but in other developing countries this figure may be underestimated (Serra-Majem L., 2001, p.674).

A FBS presents a comprehensive picture of the pattern of a country's food supply during a specified reference period. The food balance sheet shows for each food item – that is each primary commodity and a number of processed commodities potentially available for human consumption – the sources of supply and its utilization. The total quantity of foodstuffs produced in a country added to the total quantity imported and adjusted to any change in stocks that may have occurred since the beginning of the reference period gives the supply available during that period. On the utilization side a distinction is made between the quantities exported, fed to livestock, used for seed, put to manufacture for food use and other uses, losses during storage and transportation, and food supplies available for human consumption.

The per caput supply of each such food item available for human consumption is then obtained by dividing the respective quantity by the related data on the population actually partaking of it.

LITERATURE REVIEW AND OBJECTIVES

It is important to note that the quantities of food available for human consumption, as estimated in the food balance sheet, relate simply to the quantities of reaching the consumer.

However, the amount of food actually consumed may be lower than the quantity shown in the food balance sheet depending on the degree of losses of edible food and nutrients in the households, e.g. during storage, in preparation and cooking, as plate-waste or quantities fed to domestic animals and pets, or thrown away.

FBS do not give any indication of the differences that may exist in the diet consumed by different population groups, e.g. different socio-economic groups, ecological zones and geographical areas within a country; neither do they provide information on seasonal variations in the total food supply.

The accuracy of FBS, which are in essence derived statistics, is of course dependent on the reliability of the underlying basic statistics of population, supply and utilization of foods and of their nutritive value. These vary remarkably between countries, both in terms of coverage as well as in accuracy.

Countries with no routine information on the food consumption of their population and those interested in comparing their national dietary patterns with those countries have traditionally used the FBS assembled by the FAO (Naska A. et al., 2001b, p.1159).

Despite their limitations, FBS are useful in that they indicate the (in)adequate aspects of food supply and give crude of (un)desirable changes in terms of potentials (adverse) expected health impact. As a result of their long history, FBS are especially used for assessing trends over time (Kroes R. et al., 2002, p.343-344).

2.2. AVAILABILITY

Food availability is a major factor influencing food choice since, *a priori*, food choice can only be made from the foods available for purchase, exchange or obtainable from the environment.

In developing societies food availability and the related capacity to purchase the food are major determinants of food choice although these will always be modulated by cultural factors defining what are seen as proper or acceptable foods. In non-industrialised societies the availability of foods has a strong seasonal dependence and the patterns of food consumption therefore show seasonal cycles (Meiselman H.L., 1996, p.379-380). The availability of different foods probably has an influence on eating habits (Harro M. et al., 2005, p.115).

Philosophically, it is quite difficult to distinguish between the idea of a food product *not* being consumed because of, on the one hand, high price, and on the other, non-availability. When prices rise, the product drops out of the diet completely for some consumers and so, in one sense, consumption falling because of rising prices. We can make a distinction between the kind of effects of supply-oriented factors, which modify the relative prices of different foods and this alters the balance of the diet, and those cases where supply-oriented factors influence the availability of a food product in a particular country or region (Marshall D., 1995, p.34).

Food intake

We can only eat what is there to eat.

If the culture and environment limit the amount available, there is no escaping this constraint. Cultural factors limit availability in a number of ways. They co-determine, with economic/environmental factors what is grown or raised, and what is imported.

Price is a major practical determinant of what is effectively available, and hence intake. And convenience is regularly cited as an important factor in determining food choice and presumably amount eaten.

Food choice

Availability is ultimately the product of individuals and their preferences and abilities. Widespread interest in a food within a culture spurs attempts to obtain more, by local means or by importation, and to develop technologies to increase availability and lower price. High levels of desire in Europe over the last hundreds of years have caused coffee, chocolate and sugar to move from prohibitively expensive, luxury items, to commonplace parts of the daily diet.

The understanding of the effects of availability and price falls largely outside of the domain of the psychology of food choice.

Many of the influences on food choice are likely to be mediated by the beliefs and attitudes held by an individual. Beliefs about the nutritional quality and health effects of a food may be more important than the actual nutritional quality and health effects of a food may be more important than the actual nutritional quality and health consequences in determining a person's choice. Likewise various marketing, economic, social, cultural, religious or demographic factors may act through the attitudes and belief held by the person (Meiselman H.L., 1996, pp.86).

3 DAFNE – Data Food Networking

3.1. HBS HISTORY BACKGROUND

The HBS in the European Union are sample surveys of private households carried out regularly under the responsibility of the National Statistical Offices in each of the fifteen Member States (European Statistical System). Essentially, they provide detailed information about household consumption expenditures on goods and services, with considerable detail in the categories used; information on income, possession of consumer durable goods and cars; basic information on housing and many demographic and socio-economic characteristics.

Unlike other European statistical domains, the HBS is voluntary and no EU regulation exists. There is thus considerable freedom for each Member State to decide the objectives, methodology, programming and resource assignment for their respective HBS.

In co-operation with National Statistical Offices of the Member States, Eurostat has for many years worked on the quality – mainly the comparability of HBS statistics within the EU (Figure 2). In spite of the important progress already done, there is still big room for improvement regarding quality and harmonisation of HBS data (Eurostat, 2003, p.7).

In most European countries HBS are conducted regularly on a national basis, thus enabling them to be used for within- and between-country comparisons of food intake patterns. Furthermore, the availability of detailed food acquisition data and hence the low cost of data collection for epidemiological analyses is a major advantage compared with other methods of obtaining nutritional data. Thus, HBS can be used as an important tool for monitoring nutritional intakes in European countries (Paterakis S.E. and Nelson M., 2003, p.571).

The latest HBS in the European Statistical Member States took place between 1998 and 2000 (depending on the country). Eurostat centred the survey years on the common reference year 1999.The new applied classification of goods and services items (COICOP-HBS 1999) is presented in ANNEX 5 together with explanatory notes (Eurostat, 2003, p.8-10).

Belgium:	1999	Enquête sur les udgets des Ménages
Germany:	1998	Einkommens- und Verbrauchsstichprobe
Greece:	November 1998 – October 1999	Family Budget Survey
Spain:	1998	Encuesta Continua de Presupuestos Familiares
Ireland:	June 1999 – July 2000	Household Budget Survey
Austria:	November 1999 – October 2000	Konsumerhebung
Portugal:	Janary 2000 – January 2001	Inquérito aos orçamentos familiares
Finland:	1998	Kulutustutkimus
United Kingdom:	April 1999 – March 2000	Family Expenditure Survey

Figure 1. Timing of the national surveys (Eurostat, 2003, p. 12)

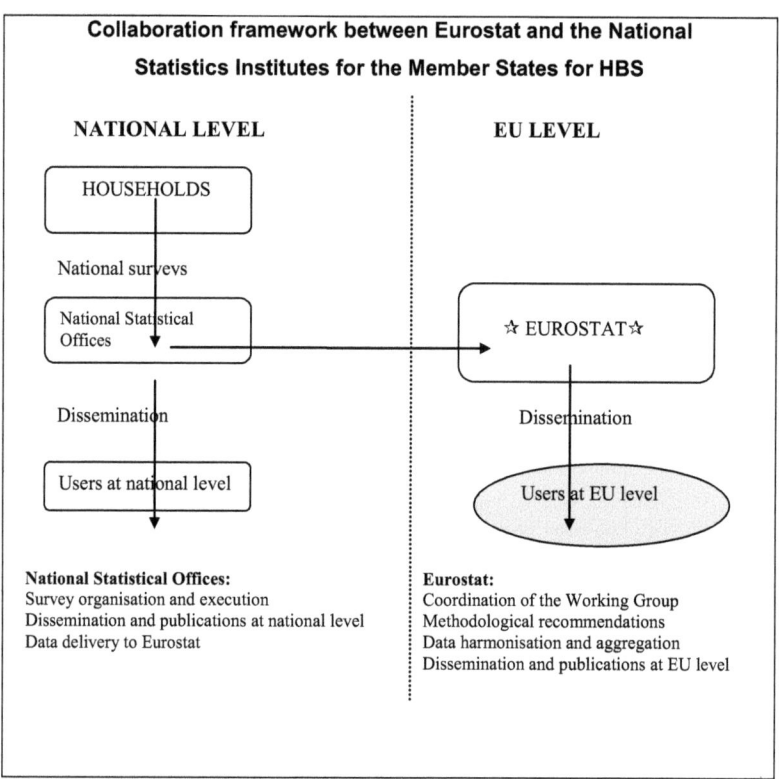

Figure 2. Collaboration framework for HBS (Eurostat, 2003, p. 12)

Historically, the prime objective of conducting HBS in all the Member States of the European Statistical System was to collect information on household consumption expenditures for use in updating the 'weights' for the basket of goods used in the Consumer Price Indices (CPI). Over the years, the range of uses has grown, as the surveys also had to meet the requirement to give a picture of living conditions of private households in certain areas and periods of time. Hence HBS are multi-purpose surveys which supply for a large number of uses and users. In terms of the complexity and detail of information supplied, the surveys are an invaluable source on economic and social living conditions of households and individuals in the Member States of the European Union.

These might be studies focusing on certain subjects for example patterns of consumption expenditures or income, or relations between different subjects for example patterns of consumption expenditures or income, or relations between different subjects such as the influence of the level of income on consumption patterns, or patterns of consumption expenditure in relation to different types of households. Further, analyses of the variations in levels of living over a period of years are applied and disparities among households in the different socio-economic groups, geographical areas, rural and urban zones can be observed (Eurostat, 2003, p.8-10).

3.2. AUSTRIAN HOUSEHOLD BUDGET SURVEY - KONSUMERHEBUNG

The Austrian Household Budget Survey (HBS) is the only statistical survey, which collects data about expenses and income of private households. Additionally the HBS records data on the quantity of food and beverages acquired at household level. The Austrian HBS records all purchases, own production and contribution in kind.

The nationally representative HBS provides a specific database of extensive household data in combination with socio-demographic characteristics.

In Austria HBS have a long tradition. In 1916 the Department of Commerce published a survey about the economical status of Viennese working-class-families. Since 1945 HBS have been carried out every 10 years.: 1954/55, 1964, 1974/75, 1984/85, 1993/94 and 1999/2000.

It has to be noted that the methodology was improved in the course of time. In 1974 for the first time all types of households (yore only urban and rustically households) and equivalence scale was applied. Since 1984 the subsequently assessment/valuation has taken place of sampling based on consumer prices. In1993/94 the insertions of fictitiously rent values were included (Statistik Austria, 2004, p.4).

The Austrian data provided to the DAFNE IV project, are based on HBS carried out in 1999/2000. A specific food classification system was developed, in order to provide data comparable between countries. The DAFNE food classification system and socio-economic categories were applied to Austrian HBS data and reproduced valid estimations of food availability concerning food habits of the entire Austrian population and of population sub-groups defined on the basis of their socio-economic characteristics. The DAFNE databank allows international comparisons and monitoring of daily food availability.

In 1999 the last survey was adapted to European standards, among others to the international COICOP *(Classification of Individual Consumption Expenditures by Purpose)* nomenclature (version 1997).

The previous HBS data are not comparable with the results of HBS 1999/2000 due to the improvements of the methodology (Statistik Austria, 2003a, p.9).

3.3. DAFNE HISTORY

The ability to monitor and compare the dietary habits of different populations is important in the development of nutrition-related scientific hypotheses, in the formulation of dietary recommendations and in planning and implementing national food, nutrition and agricultural policies.
The European Commission emphasised the importance of standardised and comparable dietary data and the promotion of nutrition surveillance systems in Europe.

Within this context, the DAFNE initiative has for the last 10 years been supported by the European Commission in order to develop a European nutrition-monitoring system, based on data collected through the nationally representative HBS. The DAFNE initiative has been successful in creating a cost-effective databank of currently 16 European countries (Trichopoulou A. et al., 2005, p.69).

In Europe, there is a need for dietary assessment tool that would provide a continuous and comparable flow of information. A series of projects has been implemented aimed at the development of a cost –efficient way of using food and related data already collected in HBS. These data would provide valuable complementary information to that derived from INS (Naska A. et al., 2001b, p.1160).

The starting point for the DAFNE initiative were two workshops funded in the late 1980s by the World Health Organisation and the European Union which explored the possibility of obtaining harmonised and comparable nutrition information from various European countries using HBS data (Breslin L., 2001, p.1129).

Since 1987, the National Nutrition Centre in Athens, Greece, has organized a series of workshops, seminars, and pilot research projects aiming at the development of the most appropriate way of using food and related data from HBS. In 1993-1994 the DAFNE I and II projects provided a food database with comparable HBS data from 10 European countries. The projects were financial supported by the European Commission.

After the development of the DAFNE databank the work undertaken in the context of the DAFNE initiative was aiming at comparing data from HBS. Thus, the DAFNE databank can serve as a tool for the identification of disparities in food habits among European countries and their socio-demographic determinants.

3.3.1. DAFNE I

In 1993, the approach aiming at the exploitation of HBS data for the assessment of nutritional information was granted support of the European Union through the "Cooperation in Science and Technology with Central and Eastern European Countries" program of the Commission of the European Communities.

Thus the Scientific Network for Pan-European Food Data Bank based on HBS, in short DAFNE was created. Its tasks included the study of current methods of HBS data collection and processing, as well as the resolution of all scientific and technical issues for the consolidation of the national data bases. The end product of this effort is the creation of an operational European HBS Food Data Bank, accessible to all.

The main objectives of the DAFNE can be summarized in the following:

- the creation of a coherent infrastructure for nutrition information
  the achievement of integration, modularity and standardization
- in equipment, software and terminology
  the establishment of methods and ways enabling nutrition
- information of all types to be shared, combined and compared
  the dissemination of related research results to other European and non – European countries.

In this context the project aimed at the development of a HBS food database which would permit all interested bodies to have access to the data collected in each country separately and in all countries at the same time, in order to identify differences concerning dietary patterns and high risk population groups on account of their nutritional habits.

Five countries participated in the DAFNE I project:

Belgium, Germany, Greece, Hungary, Poland (Trichopoulou A. and Lagiou P., 1997a, p.3)

3.3.2. DAFNE II

In November 1994, the approach aiming at the exploitation of HBS data for the assessment of nutritional information was, for a second time, granted support of the European Union through the "Agriculture and Agro-Industry including Fisheries" program of the Commission of the European Communities.

The experience in methodology and software gained from DAFNE I were a great help to the DAFNE II project and speeded up processes in many aspects of the work of the project.

The main objectives of DAFNE II were phrased as:

i) the elaboration of a coherent method for comparative nutrition information retrieval,
ii) the achievement of integration, modularity and standardization in methods and terminology, enabling nutrition information of different types to be shared, combined and compared.

Six European countries participated in the DAFNE II project: Greece, Ireland, Luxembourg, Norway, Spain, the United Kingdom (Trichopoulou A. and Lagiou P., 1999, p.8-9)

3.3.3. DAFNE III

In December 1999 the DAFNE III project started.

The "European Food Availability Databank based on HBS- DAFNE III" project was aiming at updating the DAFNE databank, by including additional datasets from six already participating European countries (Belgium, Greece, Norway, the Republic of Ireland, Spain and the United Kingdom) and also by integrating several HBS datasets (collected in consequent time periods in each country) from France, Italy and Portugal (DAFNE III team, 1999, p.6-7).

The DAFNE III project was supported by the Health Monitoring Programme of DG-SANCO. The project aimed the enlargement of the DAFNE database both in terms of partnership, France, Italy and Portugal were added to the list of DAFNE countries, and in terms of volume of information stored.

Datasets collected in different time periods in each country were added, resulting in a food database which currently includes 45 surveys that cover 20 years.

The HBS data used in the DAFNE III project cover the period of 1980 – 2000 in nine countries:

Belgium, France, Greece, Ireland, Italy, Norway, Portugal, Spain, the United Kingdom (Trichopoulou A. and Naska A., 2003b, p.5-7)

3.3.4. DAFNE IV

In October 2002 the project DAFNE IV started. The Consortium of the "European Food Availability Databank based on HBS – DAFNE IV" project.
- National HBS datasets from the following five EU Member States were assembled and harmonized: Austria, Finland, Germany, Portugal and Sweden. The newly harmonized datasets were incorporated in the DAFNE database, which now forms a bank of food data from multiple HBS undertaken in each of 16 European countries (15 EU Member States and Norway).
- Developing a methodology for estimating the daily availability of selected nutrients, using food data from the national HBS. The methodological approach was pilot tested in Greek and German data.
- A protocol for collecting was compiled, in the context of the HBS, information on meals taken outside the household, addressing thus a limitation in the use of HBS data for nutritional purposes (DAFNE IV team, 2002a, p.5-6).

Five EU Member States participated in the DAFNE IV project:
Austria, Finland, Germany, Portugal, Sweden

3.3.5. DAFNE V

Currently the DAFNE network is being extended to five new EU Member States (Cyprus, Latvia, Malta, Slovenia and the Slovak Republic) and three countries of the West Balkan region (Albania, Croatia, and Serbia and Montenegro) (Trichopoulou A., 2005, p.2).

3.3.6. DAFNE databank

For the last four years and in the context of the Health Monitoring Programme of the Directorate Genera "Health and Consumer Protection" (DG-SANCO), the European Commission has been supporting the update of the DAFNE databank with datasets from new countries and with additional datasets from the previous ones. Currently, the DAFNE databank includes harmonised and comparable information from 56 surveys of 16 European countries covering the period from 1980 to 2000.

The DAFNE databank has been recognised as a tool to follow up trends in nutritional practises; to identify population subgroups whose dietary habits are not favourable according to current scientific knowledge on the association of diet and health; and to outline preventive interventions in order to support consumer choices towards healthy nutrition (Trichopoulou A. et al., 2005, p.70).

Since HBS are not designed to primarily serve nutrition purposes, the food coding systems vary among countries ranging from very detailed records to more aggregated ones. With food data recorded at various detail levels, the aggregation of food items to the lowest level of information became necessary. In cases, however, in which more than one food item were recorded under a food code and the national representatives were in a position to define the proportional contribution of each food item, the codes were correspondingly split. To assure comparability among countries, without compromising the particular dietary characteristics of the populations, disaggregating factors were also applied in food items with particular synthesis. For example, the ingredients of margarines made of vegetable and fish oil, consumed in Scandinavia, were considered through the classification procedure. To achieve disaggregation, information was derived from various national sources, including the industry, market shares and national dietary surveys. The letter was the case in the disaggregation of the food codes used in the Italian HBS, where the data of the INN-CA study were taken into consideration.

For harmonising units of measurement and for allowing aggregation of specific items into major food groups, density and conversion factors had to be employed. For instance, to estimate the overall availability of non alcoholic beverages, the grams of coffee and tea leaves purchased by the households had to be transformed to liquid equivalents, before being added to other non alcoholic drinks. The national representatives were provided by a list of factors to apply and were asked to indicate either their approval or suggestion for a change, appropriately substantiated. Due to the importance ready-to-eat dishes are gaining in the food market and in an attempt to capture consumers` preferences in relation to meat, fish and vegetable consumption, respective subgroups were defined. Hence, the respective food codes were classified under meat or fish dishes when denoted as such. Vegetable dishes and the vegetable ingredients of other preparations were classified as processed vegetables. It should, however, be noted that the decision of the DAFNE network were often shaped to accommodate the available information. Nevertheless, the DAFNE food classification system is non-static and is designed to accommodate amendments, if future developments in the food sector impose them. Towards this direction, the HBS data compilers may consider the provision of more detailed information, in particular to certain food products and dishes.

In certain countries, data were not recorded for all the food codes included in the national codification system. Although these codes do not contribute in the daily food availability in this country, they are included in the DAFNE-classification scheme in order to demonstrate their grouping, if records were collected (EU Commission, 2005, p.4-5).

3.4. COICOP – Classification of Individual Consumption of Purpose

All HBS in the EU member countries use the COICOP food classification scheme based on the European Combined Nomenclature and transformed by EUROSTAT. Even so, the HBS classification schemes only cover commodities in broadly defined food groups (e.g. "Other cereals and preparations") and not foods as consumed (Ireland J.D., 2000, p.532).

Definition:
The classification of individual consumption by purpose is a classification used to classify both individual consumption expenditure and actual individual consumption.

Context:
- COICOP is one of the "functional" classifications designed to classify certain transactions of producers and of three institutional

sectors, namely households, general government and non-profit institutions serving households. They are described as "functional" classifications because they identify the "function" – in the sense of "purpose" or "objectives" – for which these groups of transactions engage in certain transactions (OECD, 2006).

4 MATERIAL AND METHODS

Dataset used

The requested variables from the HBS data which were asked for and provided by Austrian National Statistical office to be studied by the DAFNE team are listed below:

List of the requested variables

1. General information
   - ✓ household identification number, which is the identification of each household record
   - ✓ trimester of participation
   - ✓ recording period code, necessary for spotting the exact record period, which may differ

2. Nutritional information
   - ✓ food code
   - ✓ total food expenditure (food expenditures outside the household included)
   - ✓ expenditure per food item or group, mainly for calculating the proportion of expenses on food to the household total expenses, which is an additional indicator for the financial level.
   - ✓ amounts per food item or group, this is the most important variable, since food availability is retrieved from this parameter.

MATERIAL AND METHODS

3. Socioeconomic information
- ✓ degree of urbanization of household (urban, rural, semi-urban)
- ✓ name of geographical area where the household is situated, mainly to support the creation of comparable, among countries, categories of locality and for further analysis of differences in food choice due to geography.
- ✓ household size the number of household members is used for calculating the per capita food availability.
- ✓ household composition
- ✓ age and gender of household head and members
- ✓ relationship of household members with the household head
- ✓ household disposable income (net income)
- ✓ household total expenditure
- ✓ occupation / employment status / economic activity of household head and members to identify specific behaviour on food choice
- ✓ education of household head and members for calculating whether the level of education is influencing the food patterns
- ✓ income of household head
- ✓ medical expenses data could serve as an additional information when comparing food habits with morbidity and mortality data

(Trichopoulou A. and Lagiou P., 1999, p.20)

A large amount of information concerning nutrition is collected in almost all European countries by the statistical offices through HBS at variable intervals (Kanellou A., 1999, p.5).

MATERIAL AND METHODS

Data collection in the HBS

The members of the participating households are asked to record mainly in open questionnaires all food purchases, contributions from the household's own production and the food items offered to members as gifts. At present, within the European Union, the recording period for food acquisitions mainly varies between seven and fourteen days and data collection is accomplished within one year to capture seasonal variation in food intake. The recording of demographic and socio-economic characteristics the household members allow exploratory analyses on the evaluation of their effects on dietary choices.

HBS provides regularly updated dietary data that can be linked to socio-demographic indicators and are undertaken in nationally representative population samples. HBS is not primarily designed to collect nutritional information, the food data have limitations which need to be taken into consideration. In most cases no records are collected on the type and quantity of food items and beverages consumed outside the home. The majority of the European countries collect data only on expenses related to meals out of home, except the United Kingdom where data on the out-of-home consumption have been recorded since 1992 (Trichopoulou A. and Naska A., 2003a, p.25-28).

The exploitation of HBS-derived data for nutritional purposes has been evaluated in the context of the DAFNE initiative. The DAFNE project has demonstrated that comparisons at the international level, using food and socio-economic data from national HBS, are feasible. Investigation of the HBS food data through comparisons with INS-generated information is

MATERIAL AND METHODS

required, in order to use the HBS data confidently for food monitoring trends and inter-country comparisons (Naska A. et al., 2001b, p.1160).

Information on food:

The degree of detail varies among countries since the structure of each questionnaire is different.

Own production and meals for guests are included in the food quantities, but not the out-of-home consumption, the losses and waste of food, neither the use of vitamin and mineral supplements for all the countries. In the participating countries there is no common handling of food stocks or large food purchases (Kanellou A., 1999, p.12).

In the case of Austria, only one HBS dataset (1999-2000) was provided to the coordinating centre and was accordingly analysed. The Austrian National Statistical Office was contacted for the provision of the three most recent HBS datasets. Statistik Austria informed that methodological changes were decided and applied for the first time in the HBS of 1999-2000, limiting thus the comparability of previous datasets with this last one (DAFNE IV team, 2002b, p.8).

Austrian daily individual food availability is based on data collected in the HBS in 1999/2000. The survey year was divided into 26 periods from November 1999 to October 2000. Each period required 14 days (Statistik Austria, 1999a, p.4).

MATERIAL AND METHODS

In a housekeeping book (Fig. 3a.+Fig. 3b.) each household collected the daily private expenditures, regardless if the product was consumed during the recording period.

Food expenditures were surveyed in specified registry-fields (e.g. cheese), an alphabetic index of food and beverages was given to each households for better finding of the respective position (1300 positions). Price and quantities for several positions (e.g. pork) and respectively the unit of quantity were enlisted (Statistik Austria, 2001, p.11).

The sample unit was defined as" private household means a person living alone or a group of people living together in the same private dwelling and sharing the essentials of living. It was not necessary that household members were related".

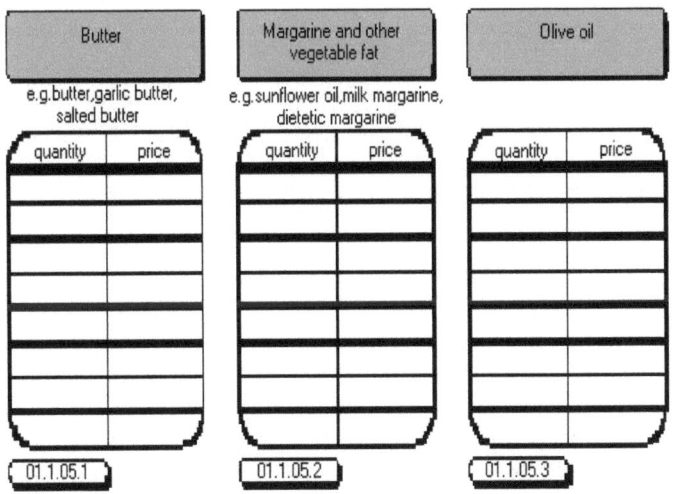

Figure 3a. Housekeeping book, Statistik Austria, 1999b

MATERIAL AND METHODS

Housekeeping book
14 days
Please fill in your food consumption!:

- White bread, crumb
  e.g.: white bread, toast, sandwich
- Dark bread
  e.g.: dark bread, grey bread, wholemeal bread
- Biscuits
  e.g.: roll, pastries
- Rusk, crispbread
- Cookies, biscuits, wafers
  e.g.: gingerbread, waffle
- Rice
  e.g.: brown rice, whole grain rice
- Pasta
  e.g.: noodles, spaghetti, dough-balls, spaetzle
- Cakes and pastries
  e.g.: cake, doughnut, strudel, flan
- Flour
  e.g.: finely ground/coarse-grained, wheat flour, wholemeal flour
- Other cereal products
  e.g.: batter, semolina, cornflakes, muesli, malt, starch
- Pork (fresh, frozen)
- Beef meat (fresh, frozen)
- Veal meat (fresh, frozen)
- Poultry (fresh, frozen)
  e.g.: chicken, turkey, duck
- Sheep and goat meat (fresh, frozen)
- Other meat (fresh, frozen)
  e.g.: game, hoarse meat
- Minced meat
- Offals
  e.g.: kidneys, tongue, entrails, brain, liver
- Other preserved or processed meat products
  e.g.: luncheon meat, corned beef
- Spread meat
  e.g.: sausage for spreading, minced pork/beef, spread of smoked pork

- Sausages
  e.g.: pork/veal sausage, frankfurter, other types of frankfurter, meat loaf
- Smoked products
  e.g.: bacon, smoked meat
- Fish (fresh, frozen)
- Seafood (fresh, frozen)
  e.g.: crabs, oysters, mussel
- Dried, smoked fish or seafood
  e.g.: smoked salmon
- Other preserved, processed fish or seafood
  e.g.: canned mussels, sardines, rollmops, herring salad
- Fresh milk
- Preserved milk
- Yoghurt
  e.g.: low-fat yoghurt, fruit yoghurt, yoghurt-drinks
- Curd cheese
  e.g.: low-fat curd cheese, fruit curd cheese, curd cheese with herbs, cheese spread
- Cheese
  e.g.: soft cheese, hard cheese, cheese sheep, mould cheese
- Other milk products
  e.g.: sour cream, cream, buttermilk, chocolate milk, crème fraîche
- Eggs
- Butter
  e.g.: butter, garlic butter, salted butter
- Margarine and other vegetable fat
  e.g.: sunflower oil, milk margarine, dietetic margarine
- Olive oil
- Other salad, cooking oil
  e.g.: table oil, pumpkin seed oil, rape oil, corn germ oil
- Animal fat
  e.g.: lard, fat of geese

MATERIAL AND METHODS

- Apples
- Pears
- Bananas
- Citrus fruit
    e.g.: Oranges, mandarins, grapefruit, lemons
- Stone fruits
    e.g. : peach, apricot, plum, cherries, coconut, avocado
- Berry fruits
    e.g.: strawberry, bramble, raspberry, redcurrant, grapes
- Other fruits
    e.g.: kiwi, melon, pineapple, mango
- Dried fruits, nuts
    e.g.: dried fruit, dried pears, sweet chestnut nuts, raisins
- Preserved, frozen fruits
    e.g.: candied fruits, salted groundnuts, frozen fruits
- Stem vegetables
    e.g.: green beans, tomatoes, cucumber, paprika, pumpkin, onion, zucchini
- Leafy vegetables, herbs
    e.g.: salad, herbs, spinach, celeriac, fennel
- Cabbage
    e.g.: cabbage, kohlrabi, herb, cauliflower, broccoli
- Root vegetables, mushrooms
    e.g.: onion, carrots, leek, asparagus, (bouquet garni) herbs and vegetables for making soup, radish
- Dried vegetables
    e.g.: dried herbs, lentil, beans
- Preserved, frozen vegetables
    e.g.: frozen food, frozen herbs, canned vegetables, gherkin
- Potatoes
    e.g.: early potatoes, late potato
- Tuber plant and other products of tuber plants
    e.g.: potato starch, mashed potato, potato dough, potato chips, (semi)-finished potato dumpling
- Sugar, sweetener
    e.g.: refined sugar in crystals, cube sugar, icing sugar, unrefined sugar
- Ice-cream
- Other confectionery
    e.g.: Turkish honey, candyfloss, Halva, Turkish delight

- Sauces, vinegar, spices
    e.g.: mustard, ketchup, mayonnaise, candied lemon peel, dressing
- Salt, spices
    e.g.: pepper, curry powder, garlic, cinnamon, chicken-grill spice
- Baking additions, soups
    e.g.: baking powder, baker's yeast, gelatine, poppy, jelly, custard powder
- Baby food
- Ready-made frozen food
    e.g.: ready frozen pizza, dumpling of yeast, apricot dumpling, ready meal
- Other preserved ready-made food
    e.g.: canned goulash, stuffed paprika in tins
- Coffee
- Tea
- Cocoa
- Mineral water (without deposit)
- Lemonades (without deposit)
- Fruit juices (without deposit)
    e.g.: orange juice, juice for watering down
- Vegetable juices (without deposit)
    e.g.: carrot juice, tomato juice
- Red wine, Rosé (without deposit)
- White wine (without deposit)
- Other fruit wine (without deposit)
    e.g.: redcurrant wine, apple wine, strawberry wine
- Other alcoholic beverages based on wine
    e.g.: champagne, sparkling wine, must, sweet cider
- Beer, non-alcoholic beer (without deposit)
- Schnapps, spirits, liqueurs
- Consumption off-site in leisure time
- Consumption in restaurants
    e.g.: tavern, pizzeria, Chinese restaurant
- Consumption in cafés ,bars
    e.g.: coffee house, tea-room, bar
- Fast Food, takeaways
    e.g.: sausage stand, takeaway pizza
- Consumption off-site in canteens/university cafeteria/nursery school, meals on wheels

Figure 3b. Housekeeping book, Statistik Austria, 1999b

MATERIAL AND METHODS

The survey was based on a sample of 7098 households. All types of private households were included and institutional households were excluded.

The response rate was 40.7 % and differed seasonally and regionally. During the Christmas and the vacation season approximately 200 households participated per recording period, but in spring and in autumn more than 300 households took part in the survey.

Federal States	Number of households
Burgenland	553
Kärnten	758
Niederösterreich	967
Oberösterreich	1020
Salzburg	656
Steiermark	1004
Tirol	658
Vorarlberg	814
Wien	668
Total	**7098**

Table 1. Distribution of households participating in the 1999/2000 Austrian HBS, by federal states (Statistik Austria, 2002, p.20)

MATERIAL AND METHODS

The sample design was generally a two-staged selection procedure for each federal state, excluding Vienna and Vorarlberg, where the selection procedure was one-staged.

The sampling frame is the Mikrozensus, which is a sample collection conducted quarterly in private households by Statistik Austria. In the course of this survey, nearly 60,000 persons in Austria – approximately 6,000 of them in Vienna – are questioned. The basic survey programme, consisting of questions about demographic, household and dwelling structures as well as employment status and unemployment, is regularly extended with different special programmes (Stadt Wien, 2002, p.23).

MATERIAL AND METHODS

Reference Period	Accounting	Contributive Households
1	01.11.-14.11.	235
2	15.11.-28.11.	188
3	29.11.-12.12.	204
4	13.12.-26.12.	186
5	27.12.-09.01.	160
6	10.01.-23.01.	227
7	24.01.-06.02.	271
8	07.02.-20.02.	353
9	21.02.-05.03.	331
10	06.03.-19.03.	340
11	20.03.-02.04.	316
12	03.04.-16.04.	326
13	17.04.-30.04.	313
14	01.05.-14.05.	233
15	15.05.-28.05.	279
16	29.05.-11.06.	191
17	12.06.-25.06.	223
18	26.06.-09.07.	213
19	10.07.-23.07.	186
20	24.07.-06.08.	334
21	07.08.-20.08.	301
22	21.08.-03.09.	286
23	04.09.-17.09.	309
24	18.09.-01.10.	433
25	02.10.-15.10.	350
26	16.10.-29.10.	310

Table 2. Reference Period and the contributive households, Statistik Austria, 2004, p.22

MATERIAL AND METHODS
4.1. STATISTICAL METHODS

4.1.1. CONVERSION OF RECORDED EXPENSES INTO QUANTITIES ON THE BASIS OF NATIONAL ANNUAL PRODUCTION QUANTITIES

The households recorded the expenses for all food items and beverages. 58 food items were recorded in quantities available in the household together with the related expenses.

In case of the other 23 food items, which were recorded only in expenses, Statistik Austria recommended the methodology to convert food expenditure to quantity data from quantities of the Austrian annual production market and the recorded HBS - expenses. (Statistik Austria, 2003a, p.18)

The Austrian Statistical Office advised the following formula for the calculation:

Availability in quantity for food =

$$\frac{\text{National Annual Production market}}{12} \times \frac{7098 \text{ hds}}{3\,241\,338 \text{ hds}} \times \frac{\text{Expenses of hds}_i \text{ for food}_j}{\text{Total expenses for food}_j}$$

- 7098 (hds) households participated in the HBS in 1999/2000.
- 3 241 338 (hds) households existed in 1999-2000.

4.1.2. CONVERSION OF EXPENSES BY THE PRICE PER UNIT WEIGHT TO REPLACE ZERO VALUES FOR QUANTITIES

Not all households specified the required quantities, but the related expenses in the questionnaire. In these cases Statistik Austria recommended to convert expenses by the price per unit weight to replace the missing quantities and to avoid too low values.

As a general rule consumed quantities of food items are not identical with applied quantities. Due to spoilage, losses and other wastages, it is assumed that fewer food items are applied as consumed (Gedrich K., 1996, p.34).

Calculation of the average per capita food availability

Analyses were conducted separately for each of the participating countries (DAFNE IV team, 2002b, p.19). The food quantities available for consumption in the household were estimated without making allowances inedible parts, preparation losses, spoilage on the plate, or food offered to domestic pets and under the assumption of equal distribution of food during the survey period.

Individual food availability was estimated by the coordinating centre as a calibrated average, taking into account the size of the household, as well as the gender and the age of the household members. Based on age- and gender-specific average energy requirements retrieved from the literature,

MATERIAL AND METHODS

consumption units were estimated for nine age groups, and separately for males and females (Naska A. et al., 2005, p.2-3).

Even though the HBS does not record availability of food items by individuals, it obtains information on household composition. Using mathematical modelling it is possible to estimate availability for household members of different types. The model can be used to estimate typical food consumption by age, sex and age-sex groups (Zintzaras E. et al., 1997, p.54).

The individuals in each household were grouped into ten 5-year age bands. The number of individuals in each family belonging to these age groups and the average food availability per person were calculated for each household.

The model is based on the assumption that household food availability is the sum of the food quantities available to all household members, characterized on the basis of their age and sex during the recording period (Trichopoulou A. and Naska A., 2003a, p.28). For each of the DAFNE countries, food availability per capita per day was calculated by dividing the household availability by the products of the referent time period and the mean household size (DAFNE project, 1999, p.16). The general for formula calculating the average availability per person of a food item or food group is given by the ratio total availability divided by number of persons (Kanellou A., 1999, p.35). After indication and advice provided by the national data supplier, a weighting factor was incorporated in the formula whenever necessary to accommodate for the sampling scheme

MATERIAL AND METHODS

(DAFNE IV team, 2002b, p.19). Descriptive statistics are calculated to depict the nutritional habits of the European populations and of population sub-groups, defined on the basis of their socioeconomic characteristics (Trichopoulou A. and Naska A., 2003a, p.25).

At first the data were converted into availability per day in grams, furthermore the average per person availability of a food item or group was calculated (Kanellou A., 1999, p.35). If a sample contains information from n households, each having m_i members and the availability of a specific food item is equal to y_i, where i is 1,2,3....,n, then estimate of the availability on a per person basis (x_i) is:

$$x_i = y_i / m_i d,$$

where d is the survey period. Weighting factors w_i, where i is 1,2,3,...,n, were taken into account for each household, depending on the sampling ratio from various population strata. In the case of countries where no such factors were introduced, their value was considered as being 1 (Naska A. et al., 2000, p.550).

5 METHODOLOGY

5.1. AUSTRIAN HBS DATA COLLECTION METHODOLOGY

GENERAL INFORMATION I

Survey years	1999 – 2000
Response period	November 1999 – October 2000
Sample size (households)	7,098
Response rate %	40.7%
Sampling frame	Häuser und Wohnungszählung 1991,Mikrozensus 1994/95
Sample design	Generally two-staged selection procedure for each federal state, excluding Vienna and Vorarlberg where the selection procedure is one-staged.

GENERAL INFORMATION II

Definition of a sample unit	Private household: means a person living alone or a group of people living together in the same private dwelling and sharing the essentials of living. It is not necessary that household members are related.
One-person household included	yes
Households excluded from HBS	All types of private households included, institutional households excluded.
Handling of refusals	Weighting
Number of regions, included in the sample	9
People paid to participate	Yes 36,33 € per participating household

METHODOLOGY

SPECIFIC INFORMATION REGARDING DATA COLLECTION I

Method for assessing food availability	Haushaltsbuch ("housekeeping book")
Recording period	14 days
Capturing of seasonal variability	The survey year was divided into 26 periods from November 1999 to October 2000, each lasting for 14 days. The valuation of own-production depended on the enquiry period.
Household income data	Available
Socio demographic information	Available
Number of food groups/items included in the questionnaire	88 food groups in the housekeeping book
Number of food groups/items recorded in expenses	88
Number of food groups/ food items recorded in quantities	81

SPECIFIC INFORMATION REGARDING DATA COLLECTION II

Aggregation of food groups/items made before or after data collection	before
Number of non-aggregated food codes stored	None
Alcoholic beverages included in quantities	Yes
Alcoholic beverages included in values	Yes
Tea, coffee, stimulants included in quantities	Yes
Tea, coffee, stimulants included in values	Yes
Consumption of own production of food recorded	Yes

METHODOLOGY

SPECIFIC INFORMATION REGARDING DATA COLLECTION III

Information concerning women in pregnancy recorded	No
Information concerning women in lactation recorded	No
Meals of guests included	Yes
Handling of waste and losses within the household.	No handling
Handling of foods given to pets.	Within the COICOP-nomenclature, pet food is a separate position.
Special handling of food stocks.	No handling on the individual level. Own-production was only recorded if actually consumed.
Special handling of large food purchase.	Food that was bought by bulk purchases is a separate COICOP-position. No special handling in relation to the consumption data.
Household living conditions recorded	Yes
Information regarding use of vitamin and mineral supplements recorded	No
Food data	Yes

METHODOLOGY

OTHER INFORMATION REGARDING THE HOUSEHOLD BUDGET SURVEYS

Extent of difference between the recorded food availability in the HBS and in the National Accounts (NAC) (%)	~ -13 % (because of conceptual differences, these values cannot be compared exactly).
% proportion of meal expenditure outside home	~ 15 – 20 % (cannot be answered exactly because the expenditures in bars/restaurants/… do not allow a differentiation between food, drinks and other expenditures)
% proportion from own production	5,2%
Nutrient conversion factors.	No
Undertaking nutritional analyses	No
Frequency of HBS	Every 5 years

5.2. DAFNE TASKS

Task 1: Provision of the raw HBS data:

The respective National Statistical Offices prepare text files with detailed raw (household per household) data on foods acquisitions, medical expenses, demographic and socio-economic characteristics (a detailed list of the pre-requisites variables is included page 20-21). The participating countries are asked to provide detailed file descriptions and additional clarifications, whenever needed.

Task 2: Cleaning of the raw HBS data:

Computer experts in the DAFNE coordinating centre continue the data reading and cleaning. Data files to be used in the analysis (one file per survey per country) are prepared and stored in a preliminary database.

The national datasets and their file descriptions were sent to the DAFNE coordinating centre in Athens. Data were read and the correct reading was evaluated through multiple cross-checks and tests of data consistency by standard procedures including:
- the assessment of agreement between overlapping variables (e.g. age given in discrete years versus age categorized in age groups)
- the cross-tabulation of variables with related content (e.g. the age of the household head compared to his/her employment status) and
- the identification of missing data for variables considered in the analysis (Trichopoulou A. and Naska A., 2003a, p.25).

Task 3: Harmonization of HBS data:

The participating centres undertake the task of classifying the food information collected in their national HBS according to the agreed DAFNE food classification system. Nutritionists from the participating countries group the four socio-demographic DAFNE variables according to the DAFNE classification schemes.

METHODOLOGY

Task 4: Calculation of the average daily food availability:

Average per capita per day food availability are calculated. Calculations are founded upon an assumed equitable distribution of food availability within the household.

Moreover, these calculations are not adjusted for the proportion of edible food available within the household. Furthermore, daily individual food availability is calculated with respect to the four socio-demographic variables under study.

Tasks 5: Preparation of national reports:

The coordinating centre provides country-specific estimates on food availability.

Results are also provided on the food availability of population segments, defined according to the four aforementioned socio-demographic variables. Participants undertake the responsibility of preparing their national report concerning trends in food availability and socio-demographic disparities in food habits in their countries based upon the DAFNE results.

Task 6: Integration of data into nutrition-related information systems:

Upon completion of task 4, the derived data are incorporated into the fully operating DAFNE databank, which consequently includes data on the food availability of 12 EU Member States and Norway during the past twenty years. The DAFNE databank is integrated into the DAFNESoft program, developed by the Coordinating centre in Athens (Trichopoulou A., 2002, p.5-6).

METHODOLOGY

Integrating DAFNE data into nutrition-related information systems

Upon completion of tasks 1-5, the derived database was incorporated into the operating DAFNE databank, which was updated to include harmonised information on the mean daily food availability in 15 EU Member States and Norway. The DAFNE databank was further integrated into the operating DafneSoft application tool (DAFNE IV team, 2002b, p.20).

DafneSoft

The DAFNE data on mean food availability (g or ml/person per day) are integrated in DafneSoft (version 2.1), which is a software for the Microsoft Windows operating system allowing:

a. The presentation of dietary data in various formats (tables, bars pie charts, map presentations) and at various levels of detail;
b. The follow-up of trends in food availability over time, within and between countries;
c. The study of the effect of the household's locality and of the education and occupation of the household head on the daily food choices; and
d. The export of data for further uses (Trichopoulou A. et al., 2005, p.69).

The application tool is freely available at www.nut.uoa.gr. – University of Athens (DAFNE IV team, 2002b, p.20)

METHODOLOGY

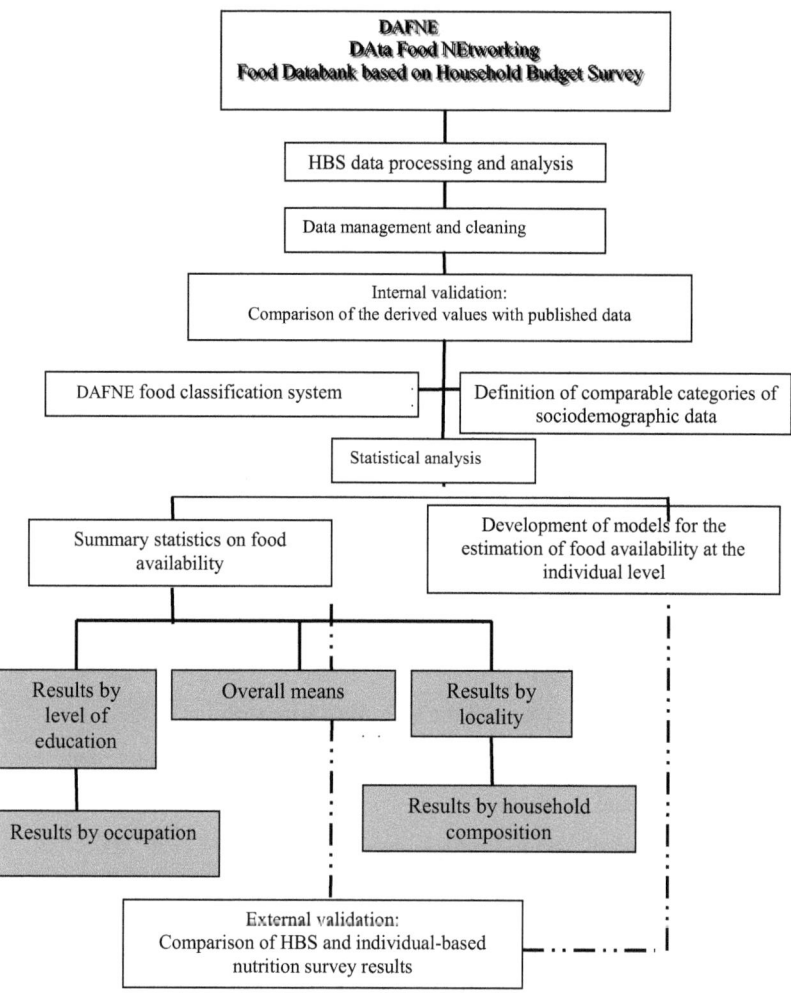

Figure 4. DAFNE methodology (Trichopoulou A. and Lagiou P., 1999, p.17)

METHODOLOGY

5.3. CREATION OF COMPARABLE TABLES REGARDING SOCIO – DEMOGRAPHIC PARAMETERS AND FOOD GROUPS AVAILABLE IN THE HBS DATASETS

HBS data from various countries are collected in a database in ASCII format.

Comparable categories are formed. Food aggregation tables and tables of socio-demographic parameters were formed (Kanellou A., 1999, p.15).

5.3.1. Definition of comparable categories of socio-demographic data, among the DAFNE-member countries

Harmonisation of the food, demographic and socio-economic information

Information derived from the DAFNE databank reveals disparities in food habits within, between and across the participating populations. Furthermore, there appears to be nutritional variations within countries by socio-demographic groups defined by residence or educational level (Trichopoulou A., 2005, p.2).

The DAFNE food classification scheme   (Table 4 – 16)

The development of a food classification system that would allow international comparison is a central element in the development of a

METHODOLOGY

European food databank. The DAFNE food classification allows the classification of the food data collected in different European countries into 56 detailed subgroups. These subgroups can be further aggregated at various levels ending up at 15 main food groups (Trichopoulou A. and Naska A., 2003b, p.7-8).

The DAFNE socio-demographic classification scheme (Figure 5 – 8)

Several socio-demographic characteristics are recorded in the HBS and many of them were included in the final roster of variable to be studied. It was decided to initially focus on locality (degree of urbanisation of the area where the household was situated),
education and occupation of the household head, and on the household's composition.

These variables were used for characterisation of the socio-economic status of the household (Trichopoulou A. and Naska A., 2003a, p.25) and were identified in each dataset and were classified under common socio-demographic groups, formed through the establishment of operational criteria. Specific problems were addressed and between countries comparability was assured through interactive cross coding, several working group meetings and bilateral visits (Trichopoulou A. and Naska A., 2003b, p.9).

METHODOLOGY

5.3.1.1. Category of locality

Three categories of locality were formed based on the data collected in the HBS: urban, semi-urban and rural (Figure 5). Various criteria were used in the different countries (from the number of inhabitants to the prevailing occupation [agricultural versus industrial] to the number of electors) and they were all applied in an attempt to depict population density and lifestyle pattern (Lagiou P. and Trichopoulou A., 2001, p.1136).

Location of household	Finland	Austria	Greece
Rural	.Rural municipalities: municipalities which not included in the below-mentioned.	Low – all remaining areas. Municipalities, which do not attain a population density more than 500 inhabitants per km^2, are conducted among "High density" areas, when they are surrounded from such areas	
Semi-urban	.Semi-urban municipalities: at least 60% but under 90% of the population lives in urban settlements and the population of the residence area is 4.000 – 14.999.	Medium – this area has got at least 50.000 inhabitants and more than 100 inhabitants per km^2 (and no more than 500inh./km^2)	N/A
Urban	Urban municipalities: at least 90% of the population lives in urban settlements or the population of the residence area is at least 15.000.	High this area(which is a municipality or a group of municipalities, which border on each other) has got at least 50.000 inhabitants and more than 500 inhabitants per km^2	

Figure 5. Definition of urbanisation

METHODOLOGY

Two exceptions were made, based on the available HBS data:

In the case of Ireland, it was decided that the only possible categorization of the households based the data collected by the National Statistical Office would be urban versus rural, while in Luxembourg it was not possible to characterize the various cantons based on these three locality categories.

Questions have been raised as to whether the three locality categories really represent different living conditions in the participating countries. The issues of degree of urbanization need to be further studied, although the DAFNE participants agree that the existing categories reflect, more or less, the differentiation in access to facilities such as supermarkets as well as to health information messages (Trichopoulou A.. and Lagiou P, 1999, p.21).

The Austrian classification of locality is defined on the basis of **population density** (Elmadfa I. and Suchomel A., 2002).

5.3.1.2. Education of household head

Despite the differences in the educational systems of the participating countries, forming comparable categories in terms of education did not present any major problems (Kanellou A., 1999, p.16).

The educational levels of the household heads are assigned following a combination of variables reflecting years of schooling and educational attainment (Figure 6) (Elmadfa I. and Suchomel A., 2002). The educational level has an important influence on socio-economic status. Higher levels of education may also increase the ability to obtain or to understand health –

related information in general or dietary information in particular needed to develop health – promoting behaviours and beliefs in the field of food habits (De Irala-Estevez J. et al., 2000, p.709).

It was possible to form three comparable categories of educational levels.

Education of house-hold head	Finland	Austria	Greece
Illiterate/ Elementary	. Primary education or no education completed in the register	. Compulsory school attendance not completed	Illiterate Elementary education, not completed Elementary education, completed
Secondary	.Secondary education	. Compulsory school attendance completed .Compulsory school attendance and training of apprentices (vocational school) completed .Technical school . Grammar school . Secondary technical school . Secondary technical school, course for university entrance qualification/school leaving exam	.Compulsory education, completed .Secondary general education, completed .Secondary technical education, completed
Higher	.Higher education, lowest level .Higher education, undergraduate .Higher education, graduate level .Higher education, postgraduate level	. High education studies .University / College	.College/Higher education studies, completed .University/Postgraduate studies, completed

Figure 6. Definition of education

METHODOLOGY
5.3.1.3. Occupation of household head

Food choice is likely to be influenced by a number of socio-economic characteristics of the households including education, occupation, income, age etc. Not all of the DAFNE countries provide education or income data, while occupation is given for all member countries and potentially provides a common basis for comparison (Trichopoulou A. and Lagiou P., 1999, p.21). In addition, occupation reflects a number of characteristics including physical activity (DAFNE IV team, 2002b, p.17), education and income and (Trichopoulou A. and Lagiou P., 1999, p.21) if properly harmonized, can provide a suitable basis for comparison of dietary practises (DAFNE IV team, 2002b, p.17) and influences that cross country boundaries are less culturally determined (e.g. number of years of education may reflect different social levels between countries, while skills associated with a given occupation may be similar) (Trichopoulou A. and Lagiou P., 1999, p.21-22). Five DAFNE categories of occupation are assigned (Figure 7). The Austrian classification of household head's occupation is defined on the basis of the job description and the present participation in gainful employment (Elmadfa I. and Suchomel A., 2002).

METHODOLOGY

Occupation of household head						
	Finland	Austria			Greece	
		Job description	Present participations in gainful employment		Primary variable: Profession	Secondary variable: Occupational status
Manual	.Self-employed in agriculture .Blue-collar workers	.Soldiers .Qualified employees in agriculture and fishery .Professions in winning minerals and building .Metalworker, mechanics and related professions . Precision workers, art craftsmen , printers and related professions . Other skilled trades and related professions . Operators stationary and related machines . Operators of machines and fitters . Drivers of a vehicle and operators of mobile machines .Unskilled agricultural fishery and related workers . Unskilled workers of mining ,trade of building , processing and transport	.Employed .Military service/Alter native service		Farmers, stock breeders, forest guards, fishermen, hunters Craft and related trade workers (excluding workers in agriculture) and transport vehicle operators	Working

Definition of occupation

METHODOLOGY

	Finland	Austria		Greece	
Non-manual	.Other self-employed .Upper white collar .Lower white collar	. Members of legislative bodies and executive administration employees . Managers and managing director in big enterprises . Managers in small enterprises . Physicists, mathematicians, engineer scientists . Biological scientists and medics . Scientific teachers . Other scientists and related professions . Technical qualified employees . Qualified employees of biological science and health . Non-scientific teachers . Other qualified employees . Office workers without customer service .Office workers with customer service . professions in the services sector referring to persons and security men . Models, salespersons and persons, who present something .Unskilled workers of services	.Employed .Military service/Alternative service	Scientists, self-employed professionals and technical assistants Senior administrative staff and managers of private and public sectors Office clerks Tradesmen and salesmen Service workers, including military	Working

Definition of occupation

METHODOLOGY

	Finland		Austria		Greece
Retired	Pensioners		.Retired		Retired
Unemployed	.Unemployed		Unemployed		
Others	.Students .Other economically inactive (homemakers, not elsewhere classified) .Socio-economic status unknown	.Never been gainful employment	.Military service/Alternative service .Waiting period parents .Working only in the household .Pupils/Students .Other supported persons	Others, unable to be classified by profession	Student or soldier Searching for employment Household work Other

Figure 7. Definition of occupation

This scheme slightly deviates from the one developed and used in the context of the previous DAFNE III project (manual; non manual; retired; and, others). The DAFNE participants noted that the last composite category presented several problems, as it comprises individuals with different activity levels and varying incomes. Hence, DAFNE IV participants decided on a disaggregation, by separating the vulnerable group of unemployed household heads.

Since all the remaining households (students, housewives or invalid persons as households` heads) were classified in the last heterogeneous group, it was decided to label this group as "Others". If this category corresponds to more than 5% of the total survey sample, details will be thought with respect to the type of households classified under this category (DAFNE IV team, 2002b, p.17-18).

METHODOLOGY

5.3.1.4. Category of household composition

Though not frequently included in dietary analyses, household composition can have a considerable influence on nutritional choices. With children defined as being up to 18 years old, adults being between 19 and 65 years of age and individuals more than 65 years old considered as elderly, and with an intended distinction between single and other households, the following eight categories were formed (Figure 8):

Children were defined as up to 18 years of age, adults from 19-65 years of age and elderly as more than 65 years old (Elmadfa I. and Suchomel A., 2002).

Household composition of household	Finland Austria Greece	
	Number of household members	Age of household members
Single adult households	1	Member of 18-65 years of age
Households of two adult residents	2	Member of 18-65 years of age
Households of one adult resident and children (lone parent)	≥ 2	Only one member 18-65 years old, and all other members aged 0-17 years
Households of two adult residents and children	≥ 3	At least 2 members 18-65 years old, and all other members aged 0-17 years
Households of adult and elderly residents	≥ 2	At least 1 member 18-65 years old, at least 1 member of more than 66 years of age and no children (0-17 years)
Households of children, adult and elderly residents	≥ 3	At least 1 member 18-65 years old, at least 1 member of more than 66 years of age
Single elderly households	1	Member of more than 66 years of age
Households of two elderly residents	2	Member of more than 66 years of age

Figure 8. Definition of composition of household

METHODOLOGY

Because of the small number of households classified in some of the above groups, participants of DAFNE IV decided that data on the daily food availability will be reported only if:

- The group corresponds to at least 0.5% of the total survey sample AND
- More than 100 households are classified under this group.

Lastly, one group will be added including all households that are left unclassified and will be labelled as "Others".

Measures of socioeconomic position classify individuals in groups of similar status or prestige power, power, knowledge and resources. Educational, occupational and income level are used to characterise socioeconomic groups. For historical and cultural reasons, Europe traditionally uses education and occupation while for instance the US mainly relies on income and education (Galobardes B. et al., 2001, p.334).

5.3.2. Definition of comparable categories of food items among North-, Central- and South European countries

With respect to the food data collected by the HBS the first step was to clarify what exactly was included under each code used by the statistical offices. The levels on detailed recorded food data varied from one country to another. The detailed description of each food category code, provided by each national statistical office, were studied and the limitations were detected. Furthermore, in certain cases, food groups appeared to overlap

METHODOLOGY

among countries. Since the data did not share the same degree of detail, aggregation of the food items to the lowest level of information was necessary.

When more than one food item was included under one food code and information from official sources such as statistical offices, ministries or associations were available, the percentage of each of the food items in the code could be defined and the code was split (Table 3). The proportion factors are denoted in the notes underneath the tables (Kanellou A., 1999, p.19).

Minced meat	PORK	BEEF
Austria (1999/2000)	50%	50%
Finland (1998)	25%	75%
Belgium (1987/88)	70%	30%

Table 3. Minced Meat

Cereals and its products are listed in Table 4.

METHODOLOGY

Table 4. Comparable groups of cereals and cereal products

FINLAND 1998	AUSTRIA 1999-2000	GREECE 1998-1999
\multicolumn{3}{|c|}{National food code and aggregated food groups}		
BREAD AND ROLLS		
M0111202 Rye bread M0111203 Wheat bread M0111204 Other bread M0111205 Bread (not elsewhere classified)	111210 White bread 111220 Dark bread	011121 Bread (all types)
BAKERY PRODUCTS (bread and rolls excluded)		
M0111201 Crisp bread and rye crackers M0111206 Rusks and bagels M0111207 Biscuits and wafers M0111208 Taco shells and tortillas, etc. M0111501 Rice, potato and carrot pastries M0111503 Ready-to-eat sandwiches M0111601 Sweet bun loaf M0111602*70 Danish pastries and buns, etc. M0111603*70 Doughnuts M0111604 French pastries, cakes and sweet pies M0111701 Pre-prepared dough, pizza dough, etc. M011194 Easter pudding	111230 Bisuits 111240 Rusks, crispbread 111250 Cookies, biscuits, wafers 111400 Cakes and pastries	011122 Rusks 011123 Bakery products (incl. biscuits, crackers, dough, pita bread) 011141 Pastry products (teacakes, buns, cheese pies, cheese-ham pies, pizza etc.) 011142 Confectioneries based on flour and other cereal products
RICE, CEREALS AND PRODUCTS (flour and pasta excluded)		
M0111101 Rice, rice flakes and rice flour M0111103 Other rice products M0111809 Rye groats, flakes and grains M0111811 Wheat flakes, germs, grain and bran M0111812 Other groats, flakes and grains M0111813 Corn flakes, other ready-to-eat breakfast cereals M0111814 Muesli and other grain-fruit mixtures M0111815 Pop corn and other snacks of grains M0111807 Oat groats, flakes and grains M0111810 Barley groats, flakes and grains M0111903 Other ready-made gruels and porridges	111520 Other cereal products (batter, semolina, corn flakes, muesli, malt starch) 111100 Rice	011110 Rice 011151*0,04 Cereals 011152 Other cereal based products (e.g. breakfast cereals, quacker, pop corn, corn on the knob, cereals-based savoury snacks, baby food and dietary products)

METHODOLOGY

FLOUR		
M0111801 Wheat flour M0111802 Barley flour M0111803 Rye flour M0111805 Wholemeal wheat flour M0111806 Other flours and mixed flour M0111808 Semolina	111510 Flour	011151*0,96 Flour (incl. semolina)
PASTA		
M0111301 Macaroni and spaghetti	111300 Pasta	011130 Pasta

Meat and its products are listed in Table 5.

Table 5. Comparable groups of meat, meat products and dishes

FINLAND	AUSTRIA	GREECE
National food code and aggregated food groups		
RED MEAT		
PORK MEAT (fresh and frozen)		
M0112201 Meat of swine, boneless M0112202 Pork chops M0112203 Ham, uncooked M0112204 Other meat of swine, with bone M0112205 Seasoned pork, uncooked M0112805*0.25 Minced meat M0112806*0.50 Mixed meat for Karelian stew M0112807*0.50 Meat (not elsewhere classified)	112200 Pork (fresh or frozen) 112800*0.5	011221 Fresh pork 011222 Frozen pork
BEEF, VEAL AND CALF MEAT (fresh and frozen)		
M0112101 Meat of bovine animals, boneless M0112102 Meat of bovine animals, with bone M0112103 Seasoned beef, uncooked M0112805*0.75 Minced meat M0112806*0.50 Mixed meat for Karelian stew M0112807*0.50 Meat (not elsewhere classified)	112110 Beef meat (fresh or frozen) 112120 Veal meat (fresh or frozen) 112800*0.5 Minced meat	011211 Fresh beef 011213 Frozen beef 011212 Fresh veal 011214 Frozenveal
RED MEAT OTHER THAN PORK AND BEEF (fresh and frozen)		
M0112301 Meat of sheep and goat M0112801 Meat of reindeer M0112802 Venison, other meat and game	112300 Sheep and goat meat (fresh and frozen) 112700 Other meat (fresh and frozen)	011231 Fresh lamb/kid 011232 Fresh sheep/goat 011233 Frozen lamb/kid 011234 Frozen sheep/goat 011271 Other types of meat, fresh (rabbit, pigeon, game etc.) 011272 Other types of meat, frozen

METHODOLOGY

OFFAL (fresh and frozen)		
M0112803 Liver and kidneys M0112804 Blood, tongue, bone, knuckle etc.	112900 Offal	011253 Offal
POULTRY (fresh and frozen)		
M0112401 Poultry	112400 Poultry (fresh or frozen)	011241 Fresh poultry 011242 Frozen poultry
CANNED MEAT AND MEAT PRODUCTS		
M0112502 Luncheon sausages, cold cuts M0112602 Pork grilled, smoked, cooked and cured M0112604 Poultry grilled, smoked, cured, etc. M0112701 Meat preserves M0112506 Ring sausages M0112507 Other cooking sausages M0112504 Liver pâtés and pastes M0112606 Meat in aspic M0112603 Cold cuts of poultry M0112501 Salami	112510 Sausages 112520 Smoked meat products 112610 Spread meat 112620 Other preserved or processed meat products	011251 Cold cuts, cured meat, matured meat etc. 011252 Ham and bacon 011260 Canned meat and other meat products
FINLAND	**AUSTRIA**	**GREECE**
National food code and aggregated food groups		
M0111502 Ready to eat meat turnovers, etc. M0112505 Frankfurters M0112503 Other sausages, cold cuts M0112508 Sausages (not elsewhere classified) M0112601 Ham smoked, cooked and cured M0112605 Other cured meat M0112702 Other preserved meat preparations		
MEAT DISHES		
M0111102 Liver casserole M0111402 Hamburgers, hot dogs, etc. M0111302 Noodles and lasagne with sauce M0112706 Balls and patties of poultry M0112705 Meat balls and patties of minced meat M0112709 Blood pancakes, blood sausages, etc. M0112703 Cabbage rolls M0112707 Ready-to-eat soups of meat, poultry, etc. M0111303 Meat-macaroni casserole, etc. pasta meals M0112704 Meat cabbage and		

METHODOLOGY

meat potato casserole, etc. M0112711 Other meat preparations M0112708 Salads of meat, poultry, etc. M0112710 Ready-to-eat meals of meat		

NOTES FOR MEAT, MEAT PRODUCTS AND DISHES:

AUSTRIA
Code"112800"(minced meat): 50% pork and 50% beef meat.

FINLAND
Code "M0112806" (mixed meat for Karelian stew): 50% pork and 50% beef.
Code "M0112805" (minced meat): 25% pork and 75% beef.
Code "M0112807" (meat not elsewhere classified): 50% pork and 50% beef.
Code "M0112802" (venison, other meat and game): Game largely includes red meat.

Fish and seafood are listed in Table 6.

Fruit and vegetable juices are listed in Table 12.

The sugar and its products are listed in Table 13.

Table 6. Comparable groups of fish, seafood and dishes

FINLAND	AUSTRIA	GREECE
National food code and aggregated food groups		
FISH (fresh, frozen and processed)		
M0113101 Baltic herring M0113401 Herring and Baltic herring preserves M0113102 Small whitefish M0113302 Dried and cooked cod M0113303 Smoked and grilled fish M0113103 Salmon M0113104 Rainbow trout M0113105 Other fresh fish M0113106 Saithe M0113107 Baltic herring fillets M0113108 Other fish fillets M0113301 Salted fish M0113402 Tuna fish preserves M0113403 Other fish and seafood preserves M0113109 Fish (not elsewhere	113100 Fish (fresh or frozen) 113300 Fish, seafood (dried, smoked)	011311 Fresh fish 011312 Fresh fish 011313 Fresh fish 011314 Frozen fish 011331 Cured, dried and smoked fish 011332 Cured hake 011340*0,65 Canned fish

METHODOLOGY

classified)		
SEAFOOD		
M0113201 Crayfish, shrimps, squids M0113304 Seafood cooked, smoked, etc.	113200 Seafood (fresh or frozen) 113400 Preserved o processed fish or seafood	011340*0,35 Fish roe, caviar, fish pies
		011321 Fresh seafood (incl. snails) 011322 Frozen seafood (incl. snails)
FISH DISHES		
M0113407 Ready-to-eat meals of fish and seafood M0113408 Fish and seafood soups, other fish preparations M0113405 Baltic herring casseroles, etc. M0113406 Fish and seafood salads M0113404 Fish fingers, other breaded fish products		

NOTES FOR FISH, SEAFOOD AND DISHES:
GREECE:
Code "011340" (canned fish, fish roe, caviar, fish pies): 65% canned fish and 35% fish roe, caviar and fish pies.

Eggs, milk and dairy products are listed in Table 7.

Table 7. Comparable groups of eggs, milk and milk products

FINLAND	AUSTRIA	GREECE
National food code and aggregated food groups		
EGGS		
M0114701/60 Eggs	114700 Eggs	011470 Eggs and egg products
MILK		
M0114102 Whole milk M0114101 Farm milk M0114202 Skimmed milk M0114205 Milk (not elsewhere	114100 Fresh milk 114300 Preserved milk (Ultra high temperature (UHT)-Milk)	011411 Fresh milk 011421 Low fat milk 011431*2.2 Condensed milk (sweetened and unsweetened and
FINLAND	AUSTRIA	GREECE
National food code and aggregated food groups		
classified) M0114203 Processed milk (low in lactose)		unsweetened) 011432*8 Dried milk

METHODOLOGY

M0114201 Low-fat and semi-skimmed milk		
M0114301*8 Milk powder		
CHEESE		
M0114505 Unripened cheese M0114501*1.5 Emmenthal M0114502*1.5 Edam M0114514*1.5 Cheese (not elsewhere classified) M0114504 Processed cheese M0114508*1.5 Other cheese M0114507*1.5 Roquefort, blue cheese, etc. M0114506 Cottage cheese M0114503*1.5 Cheese (rich in fat) M0114509*1.5 Grated cheese	114510 Cheese 114520 Curd cheese	011451 Soft cheese 011452*1.5 Hard cheese
Milk PRODUCTS (cheese excluded)		
M0114601 Cream, processed cream, light cream M0114605 Puddings M0114401 Curdled milk, unflavoured M0114406 Yoghurt (not elsewhere classified) M0114403 Yoghurt, unflavoured M0114511 Flavoured curd, cheese soup, etc. cheese products M0114602 Sour milk and kefir M0114402 Curdled milk, flavoured M0114405 Curdled milk (not elsewhere classified) M0114603 Curdled and sour cream, cream fraiche M0114604 Cooking cream M0114404 Flavoured and infant´s yoghurt M0118501*0.6 Ice cream sticks and cones, soft ice cream M0118502*0.6 Packaged ice cream and sorbet, ice cream cakes M0114510 Curd	114400 Yoghurt 114600 Other milk products 118500*0,6 Ice cream	011440 Yoghurt 011462 Fresh cream 011461 Chocolate milk 011850 Ice cream, sorbet 011463 Milk based beverages and deserts, yoghurt deserts etc. (incl. rice pudding)

NOTES FOR MILK AND MILK PRODUCTS:
AUSTRIA
MILK-1 Code "118500" (ice cream): 1 L of ice cream = 0.6 kg of ice cream.
GREECE
Code "011431" (condensed milk sweetened and unsweetened): Mostly unsweetened condensed milk. The sweetened condensed milk contribution is negligible.
Multiplication by 1.5 is used for the conversion of hard cheese to fresh cheese equivalents.
GREECE + FINLAND
MILK-1 All milk items (e.g. condensed milk, dried milk) are converted to fresh milk equivalents:
1 unit of condensed milk*2.2 = 1 unit of fresh milk
1 unit of dried milk*8 = 1 unit of fresh milk
FINLAND
EGGS-1 The average weight for an egg is 60g. Division by 60 gives the number of pieces of eggs.

METHODOLOGY

NOTES FOR MILK PRODUCTS:
Multiplication by 1.5 is used for the conversion of hard cheese to fresh cheese equivalents.
Code "M0114404" (flavoured and infant´s yogurt): Mostly flavoured yoghurt. The infant´s yoghurt contribution is negligible
MILK "M0118501","M0118502": 1 L of ice cream = 0.6 kg of ice cream

Added lipids are listed in Table 8.

Table 8. Comparable groups of added lipids

FINLAND	AUSTRIA	GREECE
National food code and aggregated food groups		
LIPIDS OF ANIMAL ORIGIN		
BUTTER		
M0115101 Butter M0115201*0.8 Butter-vegetable oil mixture	115100 Butter	011510 Butter
ANIMAL FAT (butter excluded)		
	115500 Animal fat	011550 Lard and other animal cooking fat
LIPIDS OF VEGETABLE ORIGIN		
VEGETABLE FAT		
MARGARINE		
M0115203 Soft-margarine M0115204 Cooking margarine M0115202 Cholesterol-lowering margarine	115200 Margarine and other vegetable fat	011521 Margarine
VEGETABLE FAT (margarine excluded)		
M0115205 Coconut and nut butter, animal fats, etc.		
VEGETABLE OILS		

METHODOLOGY

OLIVE OIL		
M0115301 Olive oil	115300 Olive oil	011530 Olive oil
SEED OILS (olive oil excluded)		
M0115201*0.2/0.9 Butter-vegetable oil mixture M0115401 Other edible oils	115400 Other salad, cooking and edible oil	011540 Vegetable oils

NOTES FOR ADDED LIPIDS:
FINLAND:
Code "M0115201" (butter-vegetable oil mixture): 80% butter and 20% vegetable oil.
1 L of oil = 0.9 kg of oil
Code "M0115205" (coconut and nut butter, animal fats etc.): Mostly coconut and nut butter. The animal fat contribution is negligible.

Potatoes and other starchy roots, pulses and nuts are listed in Table 9.
In the Austrian HBS there are no information for **pulses** and **nuts**:

Pulses have been classified under **processed vegetables**, because they cannot be separated from their food code **dried vegetables**.

In the questionnaire of HBS 1999/2000 pulses were assigned to the rubric dried vegetables. The participants recorded dried herbs, lentils, beans and others together in the same column. There is no specification about the proportion of the respective food item under this food code.

The consumption of pulses is not significant among the Austrian population.

Nuts have been classified under **processed fruits**, because they are included in the food code **dried fruits**.

In the questionnaire of HBS 1999/2000 nuts were assigned to the food code dried fruits. The participants recorded dried pears, dried fruit, sweet chestnut nuts, raisins and others together in the same column. The proportions of the respective food item under dried fruits are unknown.

In the Austrian diet nuts are of minor importance.

Table 9. Comparable groups of potatoes and other starchy roots, pulses and nuts

FINLAND	AUSTRIA	GREECE
National food code and aggregated food groups		
POTATOES AND OTHER STARCHY ROOTS (potato products included)		
M0117701 Potatoes M0117801*5 Mashed potato flakes M0117804*0.62 Potato salad M0111804 Potato flour, barley and corn starch	117700 Potatoes 117800 Tuber plant and other products of tuber plants	011770 Potatoes 011780 Sweet potatoes, potato products and other starchy products
PULSES		
M0117608 Tofu M0117501 Dried peas, beans, vegetables and root crops	Pulses are included in the food code 117500 (Dried vegetables)	011751 Beans 011752 Lentils 011753 Chick peas 011754 Other pulses
NUTS		
M0116801 Nuts and almonds	Nuts are included in the food code 116800 (Dried fruits)	011682 Nuts

The aggregation procedure for vegetables is listed in Table 10.

METHODOLOGY

Table 10. Comparable groups of vegetables

FINLAND	AUSTRIA	GREECE
\multicolumn{3}{c}{National food code and aggregated food groups}		
FRESH VEGETABLES		
GREEN LEAFY VEGETABLES		
M0117102 Lettuce M0117104 Spinach, celery and other leaf and stem vegetables	117100 Leafy vegetables, herbs	011711 Greens (endive, etc.) 011712 Lettuce 011713 Spinach 011714 Parsley, celery, dill
FINLAND	AUSTRIA	GREECE
\multicolumn{3}{c}{National food code and aggregated food groups}		
CABBAGE		
M0117201 Cabbage M0117202 Cauliflower M0117203 Broccoli, red cabbage, Brussels sprouts and other cabbages M0117101 Chinese cabbage	117200 Cabbages	011722 Cabbages 011721 Cauliflower and broccoli
TOMATOES		
M0117301 Tomatoes	117300*0.49	011738 Tomatoes
CARROTS		
M0117401 Carrots	117400*0.34 Carrots	011742 Carrots
ONIONS, GARLIC AND LEEK		
M0117405 Onion M0119201 Garlic (fresh or dried) M0117804*0.08 Potato salad	117400*0.62 Onions	011743 Fresh onion 011746 Fresh garlic 011755 Dried onions and garlic 011745 Leek
OTHER FRESH VEGETABLES		
M0117403 Swedes, turnips M0117402 Beetroots M0117302 Cucumbers M0117303 Pepper M0117404 Other root crops M0117304 Peas and beans	117300*0,51 Stem vegetables 117400*0,04 Mushrooms	011739 French beans 011732 Green peas 011734 Broad beans 011731 Cucumber 011737 Pepper 011735 Eggplants

NOTES FOR VEGETABLES:
AUSTRIA:
VEGE-1 Code "117300" (stem vegetables):49% tomatoes and 51% stem vegetables.
VEGE-2 Code "117400" (root vegetables, mushrooms): 62% onions, 34% carrots and 4% mushrooms.
VEGE-3 Code "117100" (leafy vegetables, herbs): Mostly leafy vegetables. The herb contribution is negligible.

METHODOLOGY

FINLAND	AUSTRIA	GREECE
National food code and aggregated food groups		
OTHER FRESH VEGETABLES (continued)		
M0117305 Zucchini, pumpkin, aubergine and other fruit vegetables M0117406 Champions M0117409 Vegetables (not elsewhere classified M0117407 Other mushrooms		011741 Globe artichoke 011733 Zucchini 011736 Okra 011744 Beetroots 011747 Other vegetables (mushrooms incl.)
PROCESSED VEGETABLES		
M0117601 Pickled cucumbers M0117602 Pickled beetroots M0117603 Other vegetable and root crop preserves M0117604 Vegetarian patties	117500 Dried vegetables (incl. pulses)	011763 Olives 011765 Tomato paste, canned tomatoes

FINLAND	AUSTRIA	GREECE
National food code and aggregated food groups		
M0117606 Vegetable and root crop salads M0117607 Vegetable and root crop casseroles, etc. M0117605 Ready-to-eat meals of vegetables and root crops M0117408 Frozen mixes of vegetables and root crops	117600 Preserved, frozen vegetables	011762 Canned vegetables 011764 Pickles 011761 Frozen vegetables

The aggregation procedure for fruits is presented in Table 11.

Table 11. Comparable groups of fruits

FINLAND	AUSTRIA	GREECE
National food code and aggregated food groups		
FRESH FRUITS		
APPLES		
M0116301 Apples	116300 Apples	011630 Apples
CITRUS FRUITS		
M0116101 Oranges M0116102 Mandarins M0116103 Other citrus fruit	116100 Citrus fruits	011611 Lemons 011612 Mandarins 011613 Oranges 011614 Other citrus

METHODOLOGY

		fruits
BANANAS		
M0116201 Bananas	116200 Bananas	011620 Bananas
GRAPES		
M0116601 Grapes	116600*0.31 Grapes	011662 Grapes
PLUMS		
M0116501*0.60 Plums	116500*0.29 Plums	011654 Plums
BERRIES		
M0116602*0.6 Black currants M0116603*0.6 Red and white currants M0116604*0.5 Strawberries M0116605*0.6 Other garden berries M0116606*0.6 Blueberries M0116607*0.6 Lingonberries and cranberries M0116608*0.5 Cloudberries and other wild berries	116600*0.69 Berry fruits	011663 Strawberries
APRICOTS AND PEACHES		
M0116501*0.25 Peaches	116500*0.54 Peaches, nectarines and apricots	011652 Apricots 011651 Peaches
CHERRIES AND SOUR CHERRIES		
M0116501*0.15 Cherries	116500*0.17 Cherries	011653 Sour cherries and cherries
PEARS		
M0116401 Pears	116400 Pears	011640 Pears
OTHER FRESH FRUITS		
M0116701 Kiwi M0116702 Melons M0116703 Pineapple, papaya, carambola and other fruit M0116704 Fruit (not elsewhere classified)	116700 Other fruits	011672 Watermelons 011673 Melons 011656 Loquat 011661 Figs 011655 Avocado 011671 Kiwi 011674 Other fruits
FINLAND	**AUSTRIA**	**GREECE**
National food code and aggregated food groups		
PROCESSED FRUITS		
M0116802 Raisins M0116803 Other dried fruit and berries M0116901 Fruit and berry preserves M0116609 Frozen berries and berry mixes (not elsewhere classified) M0116903 Ready-to-eat berry and fruit soups and puddings	116800 Dried fruits (incl. nuts) 116900 Preserved, frozen fruit	011681 Dried fruits 011691 Preserved fruits

NOTES FOR FRUITS:
FRUIT- Multiplication by 0.5 or 0.6 is used for the conversion of berries in litres to kg.
Code "M0116501" (peaches, plums and other stone fruits): 60% plums, 25% peaches and 15% cherries.

METHODOLOGY

Table 12. Comparable groups of fruit and vegetable juices

FINLAND	AUSTRIA	GREECE
National food code and aggregated food groups			
FRUIT JUICES		
M0122301 Juice drinks, juices and nectars M0122302 Berry and fruit squashes M0122303 Juices (not elsewhere classified)	122300 Fruit juices	012230 Fruit juices (incl. condensed fruit juices)
VEGETABLE JUICES		
M0122401 Vegetable juices	122400 Vegetable juices	012240 Vegetable juices

Table 13. Comparable groups of sugar and sugar products

FINLAND	AUSTRIA	GREECE
National food code and aggregated food groups			
SUGAR		
M0118101 Lump sugar M0118102 Granulated sugar M0118103 Fruit sugar M0118104 Other sugar	118100 Sugar, sweetener	011810 Sugar
SUGAR PRODUCTS		
M0118201 Jams and purees M0118202 Marmalades M0118203 Honey M0118301 Chocolate bars and confectionery M0118401 Sweets lozenges and other confectionery M0118402 Chewing gum M0118601 Molasses M0118503*0.6 Fruit flavoured ice lollies	118200 Jam, honey 118300 Chocolate 118400 Sweets 118600 Other confectionery	011821 Honey, syrup, glucose 011822 Marmalade, fruit jelly, compotes 011841 Candies and other sweets (incl. Chewing gums) 011842 Greek sweets, sponge cakes 011830 Chocolates, pralines 011843 Wedding and birthday cakes 011860 Other confectionery (incl. Fruit in syrup, chocolate spread with nuts, baby food and dietary products containing at least 50% cocoa)

METHODOLOGY

The aggregated groups for beverages and stimulants are listed in Tables 14 and 15.

Table 14. Comparable groups of non-alcoholic beverages

FINLAND	AUSTRIA	GREECE
National food code and aggregated food groups		
STIMULANTS		
COFFEE		
M0121101 Coffee M0121102 Instant coffee	121100 Coffee	012110 Coffee beans or ground
TEA AND SIMILAR INFUSIONS		
M0121201 Tea M0121202 Herbal tea M0121203 *0.01 Ready-to-drink tea	121200 Tea	012121 Tea 012122 Herbal teas (camomile, etc.)
COCOA		
M0121301 Cocoa, powdered chocolate	121300 Cocoa	012130 Cocoa
MINERAL WATER		
M0122101 Mineral water	122100 Mineral water	012210 Mineral water, soda
SOFT DRINKS		
M0122201 Soft drinks M0122403 Other non-alcoholic drinks	122200 Lemonades	012220 Soft drinks ("cola" type, etc.) (juices excluded)

NOTES FOR NON-ALCOHOLIC BEVERAGES:
FINLAND:
Code "M0121203" (ready-to-drink): Multiplication by 0.01 is used for the conversion of litres of tea to g of leaves.

Table 15. Comparable groups of alcoholic beverages

FINLAND	AUSTRIA	GREECE
National food code and aggregated food groups		
WINE		
M0212102 Wine M0214101*0.85 Home-made wine	212110 White wine 212120 Red wine, Rose 212130 Other fruit wine 212200 Sparkling wine, champagne, vermouth	021210 Wine 021220 Sparkling wine
BEER		
M0213101 Light beer M0213102 Medium-strength beer M0213103 Strong beer M0212201 Long drinks and other light drink mixes (4.7-5.5% alcohol) M0212101 Cider (4.7% alcohol) M0214101*0.15 Home-brewed beer kits	213100 Beer, non-alcoholic beer	021310 Beer
SPIRITS		
M0211101 Liqueur, punch, etc. M0211102 Spirits, brandy, whisky, rum etc.	211100 Schnapps, spirits, liqueurs	021110 Other alcoholic beverages (brandy, ouzo, whisky, etc.)

NOTES FOR ALCOHOLIC BEVERAGES:
FILAND:
Code "M0214101" (home-made wine and home-brewed beer kits): 85% wine and 15% beer.

The food groups for miscellaneous food are listed in Table 16.

Table 16. Comparable groups of miscellaneous and dishes

FINLAND	AUSTRIA	GREECE
National food code and aggregated food groups		
FOOD ITEMS		
M0117103 Fresh herbs M0116902 Infants' juices and purees M0114204 Infant milk formulas M0117802 Potato crisps M0117804*0.3 Potato salad M0111902 Ready-made infant gruels and porridges M0119101 Vinegar M0119102 Mustard M0119103 Ketchup M0119104 Mayonnaise, salad dressings and barbecue sauces M0119105 Gravies and sauce powders M0119202 Salt M0119203 Herbal salt M0119204 Spices M0119205 Culinary herbs M0119301 Yeast M0119302 Baking powder and baking soda M0119303 Preservatives and sweeteners, etc. M0119304 Dessert sauces, pudding powders, etc. M0119305 Meat stock cubes and dehydrated meat bouillon soups M0119306 Fish stock cubes and dehydrated fish stock soups M0119307 Dehydrated vegetable soups, vegetable stock cubes M0119410 Food products (not elsewhere classified) M0117803 French-fried potatoes M0117805 Other potato products	119100 Sauces, vinegar, spices 119200 Salt, spices 119300 Baking addition, soups 119400 Other food 1110400 Non classifiable food of bulk purchase 1110500 Non classifiable food from additional income	011910 Ready-made sauces, mustard, mayonnaise vinegar etc. 011920 Spices and condiments 011930 Ready-made soups, meat and vegetable juices, baking powder etc. 011940 Other food

METHODOLOGY

M0122402 Light beer and mead extracts M0117609 Soya and other vegetable milk and cream		

DISHES

M0119308 Meat, fish, vegetable foods for infants	1110100 Ready-mad frozen food 1110200 Other preserved ready-made food 1110300 Baby food	011692 Homogenised baby food and slimming food, based on fruits 011766 Homogenised baby food and slimming food, based on vegetables

NOTES FOR MISCELLANEOUS FOODS AND DISHES:
FINLAND:
Code "M0117804" (potato salad): 62% potatoes, 30% mayonnaise and 8% onion.
Code "M0117609" (soya and other vegetable milk and cream): Mostly soya. The other vegetable milk and cream contribution is negligible.

6 VALIDATION

6.1. COMPARISON BETWEEN AUSTRIAN RESULTS OF DAFNE – HOUSEHOLD BUDGET SURVEY, AUSTRIAN STUDY ON NUTRITIONAL STATUS AND FBS

It is possible to compare data from HBS, ASNS and FBS, without forgetting the limitations and methodological differences between these three types of Austrian data sets.

The household budget survey data show home food availability in Austria. The ASNS data – compiled by the Institute of Nutritional Sciences, University of Vienna - show the food intake in Austria. The FBS data - compiled by the Food and Agricultural Organization of the United Nations- show average food supply for human consumption in Austria.

6.1.1. FBS

Results from the Austrian FBS 1999 (FAO 1999) are higher than HBS for all food groups. The differences vary from only +1% for cereals to +330% for fish/seafood. A difference of 20% was found for eggs. Values upper than 50% were received for fruits (+58,9%)potatoes (+71%), meat (+71,9%), vegetables (+74,6%), sugar and sugar products (+68,9%), added lipids (+138%), alcoholic beverages (+144,4%), milk and milk products (+169%). (Table 17)

VALIDATION

6.1.2. Austrian Study on Nutritional Status (ASNS)

ASNS aims at the documentation and assessment of the nutritional status of different population groups in Austria. (Elmadfa 2003, p. p. 6-7)

Results from the Austrian Study on Nutritional Status 2000-2002 (adults 19-65 years, single 24-h-recall) are lower than HBS for most of the food groups with the exception of vegetables (+4,2%), fish/seafood (+104,3%) and alcoholic beverages (+19,3%).

Lower ASNS-results expressed in different proportions depending on the food group varies from only -4,7% for fruits to -52% for potatoes. Lower values were received for cereals (-5,2%), milk and milk products (-7,4%). Values upper than thirty percent were obtained for meat (-31,3%), added lipids (-33,3%), fruit and vegetable juices (-36,7%) and eggs (-40%). (Table 17)

Table 17. Comparison between DAFNE, ASNS and FBS results for Austria

Food Groups – Austria	HBS (DAFNE)[1]	Intake[2]	FBS[3]
Eggs (pieces/day/person)	0.5	0.3	0.6
Potatoes (g/day/person)	100	48	171
Pulses (g/day/person)	-****	-****	2.7
Nuts (g/day/person)	-****	-****	16
Cereals (g/day/person)	303	287	306
Milk and milk products (g/day/person)	284	263	764
Meat (g/day/person)	182	125	313
Vegetables (g/day/person)	142	148	248
Fish/Seafood (g/day/person)	9.30	19	40
Fruits (g/d/person)	192	183	305
Lipids, added (g/day/person)	42	28	100
Beverages, alcoholic (ml/day/person)	171	204	418
Beverages, non alcoholic (ml/day/person)	652	n.c. ‡	n.c. ‡
Sugar and sugar products (g/day/person)	74	n.c. ‡	125
Fruit and Vegetable juices (ml/day/person)	90	57	n.c. ‡

[1] No allowance is made for food wasted or given to pets. Individual values were also estimated, without taking into account age and sex differences of the household members

[2] Austrian Study on Nutritional Status (ASNS), adults 19-65 y., single 24-h-recall, n=2585, collected 2000-2002.

[3] http://faostat.fao.org/faostat/form (13.10.04), FBS 1999, original data were expressed in kg/year/capita; daily values were derived by multiplying with 1000 and dividing by 365.

****There is no information for pulses and nuts, since they are included in other food codes and have been classified under processed vegetables and processed fruits, respectively.

‡ n.c.: not quantifiable

6.2. CONCLUSION FOR THE VALIDATION

For the differences between these three types of Austrian data sets following explanations can be found:

- **Food wastage.** HBS data are expected to be higher than ASNS and lower than FBS, since food wastage, losses on transport, storage and processing was not taken into account in the analysis of the HBS data
- **The lack of information on food and beverages consumed outside home.**
  In HBS 1999/2000 food and beverages consumed in bars/restaurants were only in expenses, however, without differentiation of the kind of meals and beverages.
- **The lack of information on food and beverages consumed in institutional households.**
  The HBS 1999/2000 took neither the consumed quantities nor the expenditures of dishes and beverages in institutional households into account.
- **Age and sex were not considered, when individualising daily availability.**
  Equal distribution of food and beverages within the household was assumed.

Over-purchasing

The very act of being surveyed may have stimulated extra purchasing during the week and month. Purchases made in bulk or stimulated by special offers and discounts plus the need to avoid running out of non-perishables may mean that a surplus is often purchased for the larder. This coupled with the fact that many foods are available in large packets of more than one week's / month's supply may frequently result in a positive food balance within a household.

Reduced food consumption

A few housewives may have simplified the diet during the survey recording period in order to make recording easier. This does not mean that the actual level of food consumption had been reduced. The very diversity of eating habits and demands within the family makes it difficult to deviate from normal patterns of eating (Kanellou A., 1999, p.76).

Vintage Effect

This can be illustrated by reference to data from the British National Food (NFS) – the longest running continuous survey of household food consumption in the world. The food expenditure in the family life cycle, characterised by the recorded age of the housewife in the NFS, provides one of the best explanations of differences in patterns of household consumption for some foods. The typical pattern, illustrated by total food expenditure, is one of less than average consumption at early stage (young single people and young marrieds), rising to a peak when the housewife is in her fifties (the children have left the household and the disposable income is high) and then declining towards the average for pensioner households.

VALIDATION

The consumption variations according to age can be attributed to two effects.

1. the structure of the household (size, number of children, etc.), proportion of meals eaten outside the home and income of the household will be closely connected with the age of the housewife, and this will influence patterns of consumption between different products.
2. consumption habits are formed by children and young adults and they carry these habits through with them as they grow older. (This is the "vintage effect".) The food habits acquired in this way will be partly supply-orientated (Marshall D., 1995, p.65).

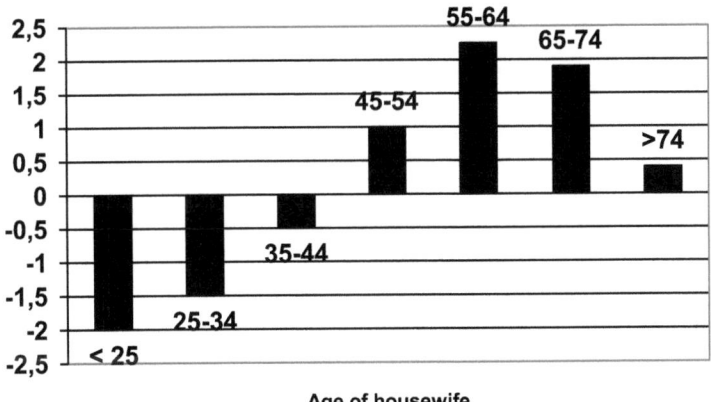

Figure 9. Total food expenditure by age of housewife (£ per person per week compared with national average) (Marshall D., 1995, p.66)

7 RESULTS – FOOD AVAILABILITY IN AUSTRIA

7.1. MEAN AVAILABILITY IN AUSTRIA

Data retrieved from the DAFNE databank, Figures - see data on p. 113

Food_Groups_AUS_1999	MEAN
Eggs (pieces/p/d)	0,50
Potatoes (g/p/d)	100
Cereals (g/p/d)	303
Milk and milk products (g/p/d)	284
Meat and meat products (g/p/d)	182
Vegetables (g/p/d)	142
Fish and seafood (g/p/d)	9,30
Fruits (g/p/d)	192
Lipids, added (g/p/d)	42
Beverages, alcoholic (ml/p/d)	171
Beverages, non alcoholic (ml/p/d)	652
Sugar and sugar products (g/p/d)	74
Fruit and vegetable juices (ml/p/d)	90

Table 18. Mean Availability in Austria, DAFNE main food groups

Data on the mean food availability of main food groups show that on the average the diets of Austrians are dominated by cereals and milk products, whereas fish and seafood are of minor importance (Table 18 and 23).

The recommended 600 grams of fruit and vegetables per person per day (DGE 2004) help to reduce the risk of some cancers, heart disease and many other chronic conditions.

The actual Austrian mean daily food availability of vegetables, fruits and juices does not attain the recommended quantity (Table 18, 23). Figure 10 shows the percentages of the daily vegetable and fruits availability. The staple foods of plant origin "cereals and potatoes" attain together a daily food availability of 403 grams per person per day.

The availability of meat and meat products (Table 23) seems to be high compared to the recommended 300-600 grams per week (DGE 2004).

In Austrian households 80% of 142g/p/d of vegetables and 95% of 192g/p/d of fruits are available fresh; Austrians present an availability of 194 ml/p/d of milk and 28 g/p/d of cheese. Austrian households show an availability of 35 g/p/d of sugar, 39 g/p/d of sugar products and 116 ml/p/d of soft drinks.

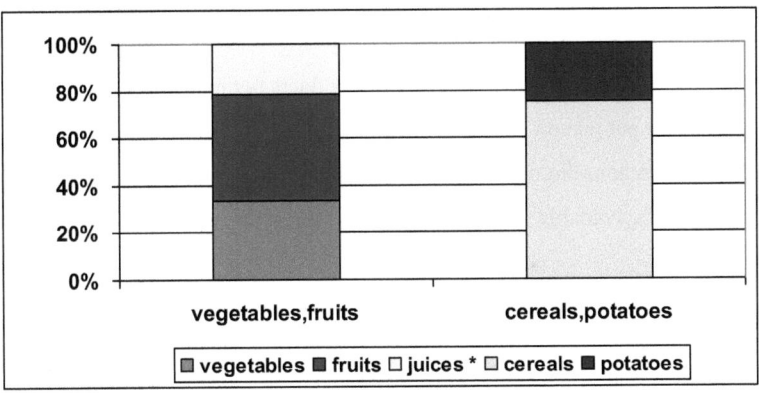

* 90ml/day ≙ 90g/day assumed

Figure 10. Mean availability of food groups of plant origin in Austria

FOOD AVAILABILITY IN AUSTRIA

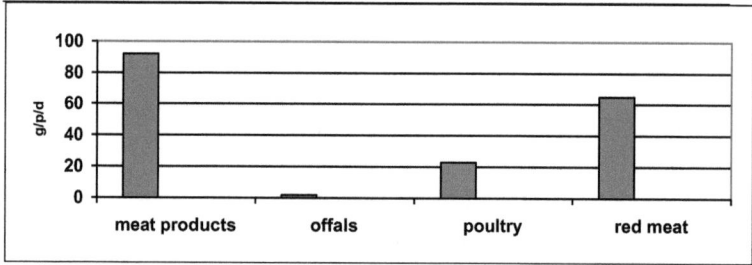

Figure 11. Mean availability of meat and meat products

Mean availability of cereals and cereal products

The households recorded an availability of 89 g/p/d of cereals and cereal products.

This compounds 75 grams other cereal products per person per day and 4 grams rice per person per day.

In Austrian households more dark bread (66 g/p/d) than white bread (25g/p/d) is available.

Austrians show an availability of bakery products of 62 g/p/d, which compounds 28g/p/d of biscuits, 18 g/p/d of cookies, biscuits and wafers, 15 g/p/d cakes and pastries and 1,21 g/p/d rusks, crispbread.

The Austrians show a mean availability of 40 g/p/d flour and of 20 g/p/d pasta.

FOOD AVAILABILITY IN AUSTRIA

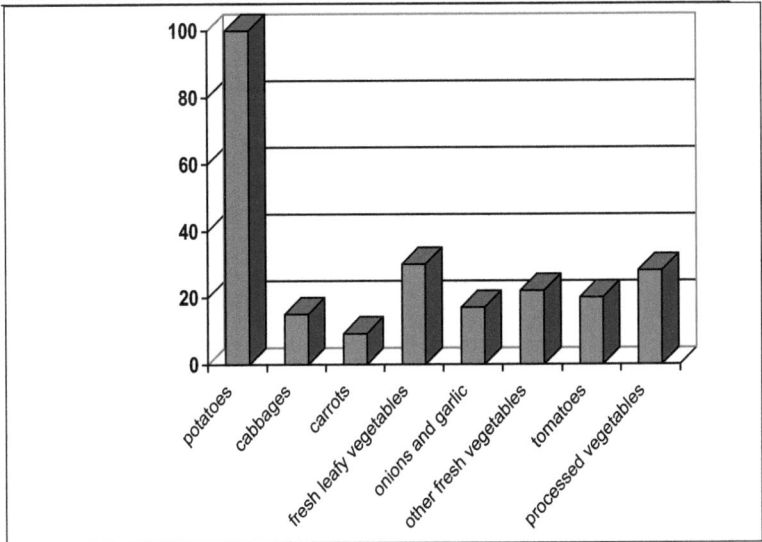

Figure 12. Mean availability of potatoes and vegetables

Mean availability of sugar and sugar products

Austrian recorded 23 grams of chocolate and 7,31 grams of sweets. 9,32 g/p/d of jam and honey were available in Austrian households. The Austrian household buys 35 g/p/d sugar, sweetener. It has to be noted that a lot of confectionery such as chocolate, chewing-gums may be bought and consumed out-of-home.

Mean availability of fruit and vegetable juices

The Austrian households recorded 88 ml fruit juices per person per day and 2,07 ml vegetable juices per person per day.

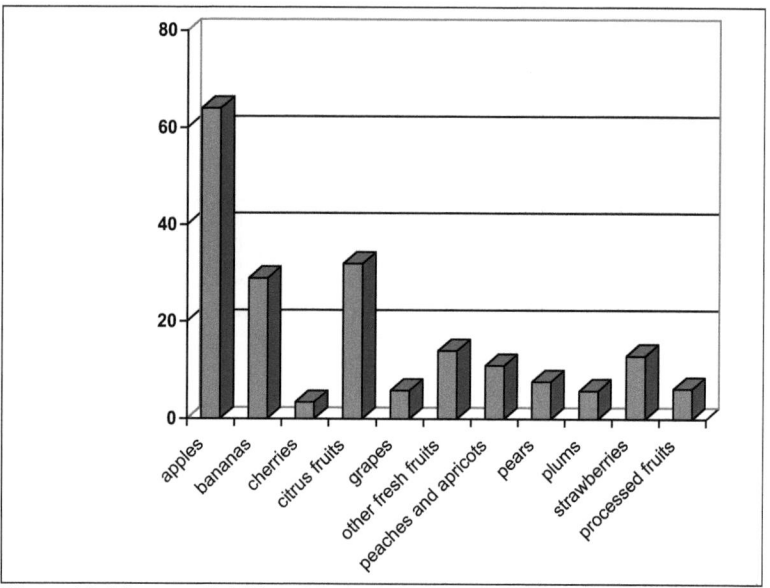

Figure 13. Mean availability of fruits (g/p/d)

The Austrian households show an availability of fruits of 192 g/p/d, which compounds 186 g/p/d fresh fruits and 6,24 g/p/d processed fruits.

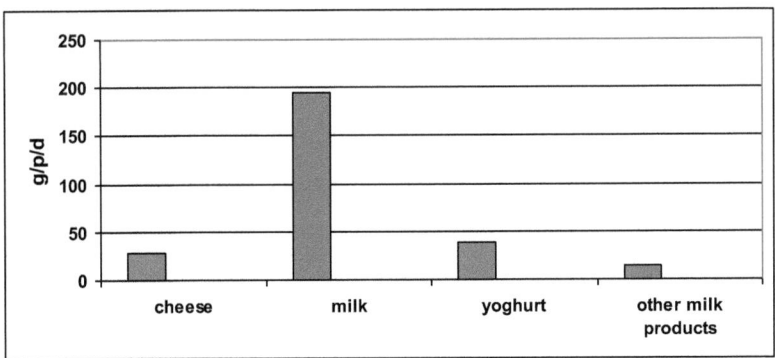

Figure 14. Mean availability of dairy products

The Austrian households show an availability of 19 g/p/d cheese and 9,52 g/p/d curd cheese. The Austrian milk availability amounts 194 g/p/d, which compounds 162 g/p/d of fresh milk and 31 g/p/d of preserved milk. Austrians show an availability of 38 g/p/d of yoghurt, 14 g/p/d of other milk products and 18 ml/p/d of ice-cream at home.

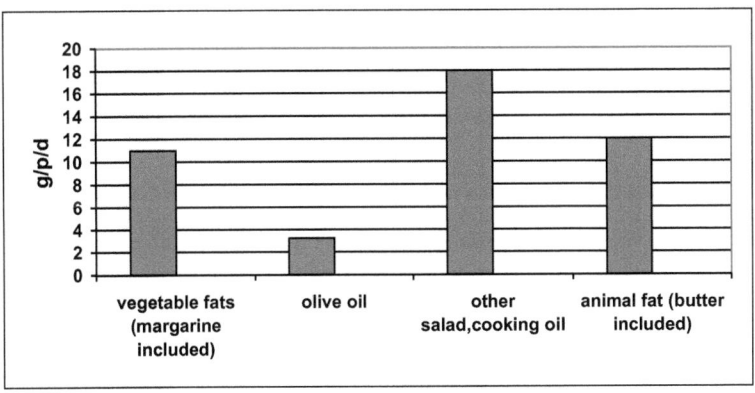

Figure 15. Mean availability of fats and oils

FOOD AVAILABILITY IN AUSTRIA

The Austrian DAFNE participants recorded a mean availability of 12 g/p/d lipids of animal origin, which compounds 11 grams butter per person per day and 1,26 grams lipids of animal origin (butter excluded).

At home the Austrian households show an availability of 30 g/p/d of vegetable lipids, which compounds 11 grams vegetable fats per person per day and 22 grams of vegetable oils per person per day. The availability of vegetable oils consists of 3,24 ml/p/d of olive oil and 18 ml/p/d of other salad, cooking oil (excluding olive oils).

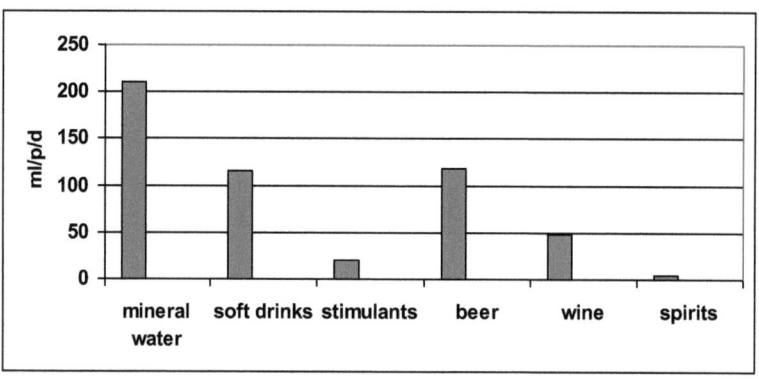

Figure 16. Mean availability of beverages and stimulants

At home Austrians drink 210 ml/p/d mineral water and 116 ml/p/d of lemonades. We can recognize an availability of 17 g/p/d coffee and 2,77 g/p/d cocoa in Austrian households.

The Austrian households recorded an availability of 119 ml/p/d beer, 48 ml/p/d wine and 4,3 ml/p/d spirit at home. It should be noted that alcoholic beverages are often drunk in restaurants.

7.2. MEAN AVAILABILITY BY EDUCATIONAL ATTAINMENT IN AUSTRIA

Data retrieved from the DAFNE databank, Figures - see data on p. A15 – A17

Education of household head _AUS_1999	1.Illiterate/ Elementary education	2.Secondary education	3.Higher education
Eggs (pieces/p/d)	0,59	0,51	0,31
Potatoes (g/p/d)	151	103	67
Cereals (g/p/d)	339	302	305
Milk and milk products (g/p/d)	343	285	267
Meat and meat products (g/p/d)	197	185	140
Vegetables (g/p/d)	193	140	159
Fish and seafood (g/p/d)	7,07	9,03	12
Fruits (g/p/d)	260	191	191
Lipids, added (g/p/d)	67	43	30
Beverages, alcoholic (ml/p/d)	172	174	142
Beverages, non alcoholic (ml/p/d)	684	660	551
Sugar and sugar products (g/p/d)	97	75	57
Fruit and vegetable juices (ml/p/d)	94	87	117

Table 19. DAFNE main food groups by education in Austria

Education of the household head

Lower educated household heads have a higher availability of **potatoes**, eggs, **cereals**, **meat**, **vegetables**, **fruits**, non alcoholic beverages, sugar and sugar products than higher educated household heads. The availability of juices, fish and seafood is higher and the availability of alcoholic beverages, **potatoes** and added **lipids** is lower among the more educated households. (Figure 17)

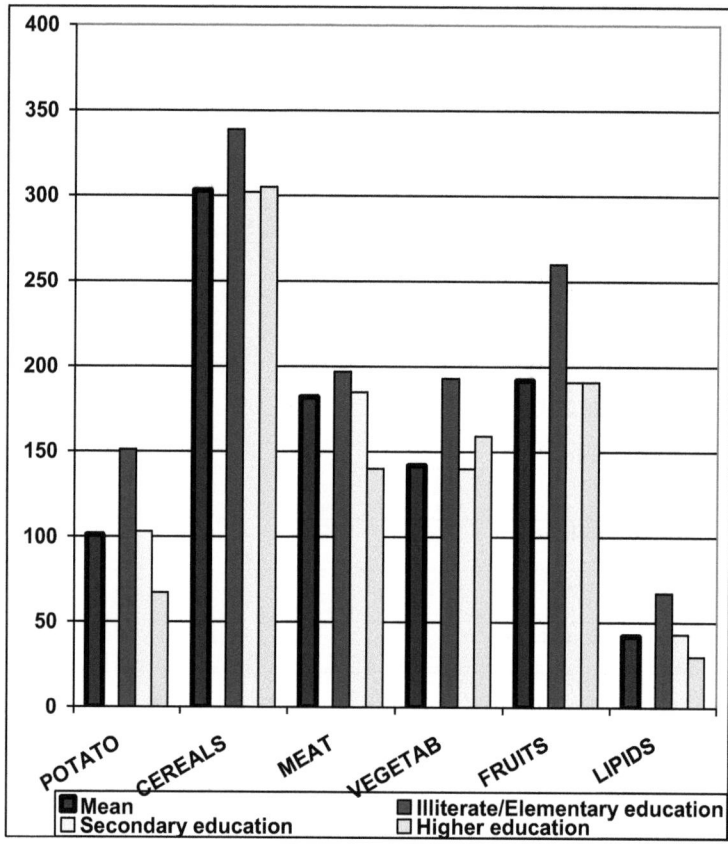

Figure 17. Average food availability, by education of the household head (g/p/d)

FOOD AVAILABILITY IN AUSTRIA

Availability of bakery products

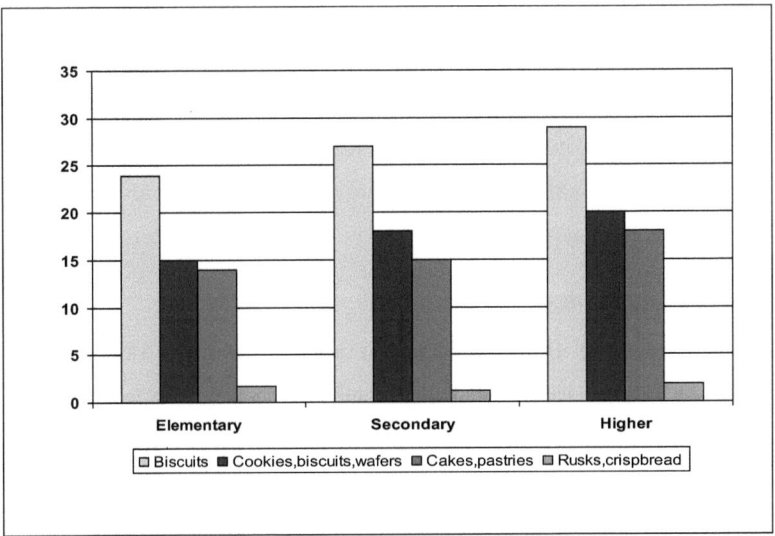

Figure 18. Biscuits, cookies, cakes, rusks by education (g/p/d)

Households with heads of lower education recorded the lowest daily availability of bakery products (55 g/p/d), whereas households with heads of higher education (68 g/p/d) recorded a higher availability value of bakery products.

Availability of bread and rolls

Households with heads of higher education show an availability of 80 g/p/d; households with heads of elementary education present a daily availability of 101 g/p/d.

FOOD AVAILABILITY IN AUSTRIA

Availability of added lipids

The highest daily availability of lipids of animal origin was recorded by "elementary educated" HH (14 g/p/d), the lowest availability was recorded by household heads of "higher educated" (11 g/p/d).

"Elementary educated" households are leaders in lipids of vegetable consumption (53 g/p/d), whereas "higher educated" households have an availability of 19 g/p/d.

Availability of offals

"Higher educated" HH show the lowest mean availability (0,88 g/p/d), whereas "elementary educated" households show the highest mean availability (1,9 g/p/d).

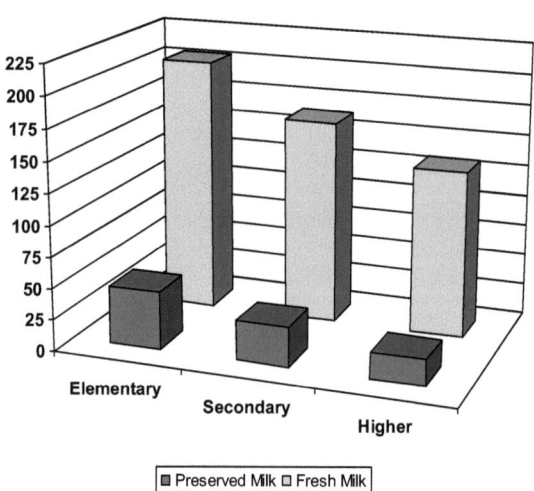

Figure 19. Fresh and preserved milk by education (ml/p/d)

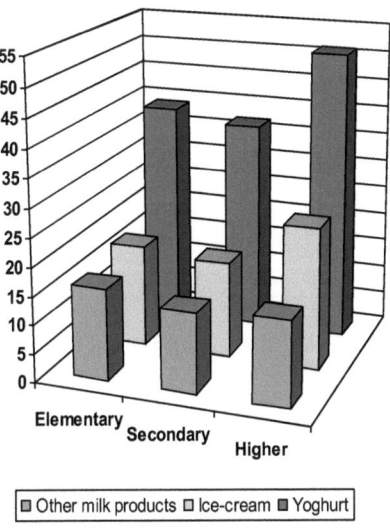

Figure 20. Yoghurt, ice-cream, other milk products by education (g/p/d)

Availability of vegetables

Fresh vegetables

The highest daily availability of fresh vegetables was recorded by elementary households (162 g/p/d), whereas the lowest mean availability was recorded by higher educated households (112 g/p/d).

Cabbage

"High educated" households recorded the lowest daily availability of cabbage (14 g/p/d), whereas "elementary educated" households recorded the highest availability (42 g/p/d).

Preserved, frozen vegetables
"Higher educated" households show an availability of 35 g/p/d, whereas "elementary" and "secondary" educated HH present a daily availability of respectively 26 g/p/d.

Availability of fruits
Fresh fruits
"Elementary educated" households recorded an availability of 255 g/p/d, "higher educated" households recorded a mean availability of 182 g/p/d.
Processed fruits
The highest daily availability was recorded by HH with heads of higher education (9,94 g/p/d). The lowest mean availability was recorded by HH with heads of elementary education (4,92 g/p/d).

Availability of beverages
Mineral water
The highest availability was recorded by "elementary educated" households (208 ml/p/d); the lowest availability was recorded by "higher educated" households (172 ml/p/d).
Lemonades
"Elementary educated" households show a mean availability of 121 ml/p/d; "higher educated" households show a mean availability of 86 ml/p/d.
Beer

"Elementary educated" households consume a daily availability of 130 ml/p/d, whereas "higher educated" households consume an availability of 86 ml/p/d.

Red wine, Rose

Households with heads of elementary education show a daily availability of 6,61 ml/p/d, whereas HH with heads of higher education present a mean availability of 20 ml/p/d.

White wine

"Higher educated" households present a mean availability of 19 ml/p/d, whereas households with heads of elementary education recorded a mean availability of 8,53 ml/p/d.

Availability of sugar, sweetener

At home households with heads of elementary education have the highest availability of sugar, sweetener (50 g/p/d). The lowest availability (15 g/p/d) was recorded by households with heads of higher education.

FOOD AVAILABILITY IN AUSTRIA

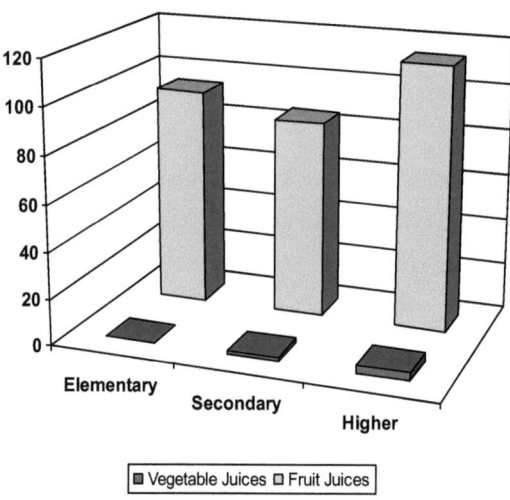

Figure 21. Fruit and vegetable juices, by education (ml/p/d)

7.3. MEAN AVAILABILITY BY LOCALITY IN AUSTRIA

Data retrieved from the DAFNE databank, Figures - see data on p. A12 – A14

Locality _AUS_1999	1.Rural	2.Semi-urban	3.Urban
Eggs (pieces/p/d)	0,56	0,50	0,42
Potatoes (g/p/d)	110	104	87
Cereals (g/p/d)	295	301	314
Milk and milk products (g/p/d)	294	268	283
Meat and meat products (g/p/d)	190	186	169
Vegetables (g/p/d)	133	147	150
Fish and seafood (g/p/d)	7,88	9,64	11
Fruits (g/p/d)	189	202	189
Lipids, added (g/p/d)	44	44	39
Beverages, alcoholic (ml/p/d)	173	200	147
Beverages, non alcoholic (ml/p/d)	630	654	675
Sugar & sugar products (g/p/d)	80	70	70
Fruit & vegetable juices (ml/p/d)	73	86	112

Table 20. DAFNE main food groups by locality in Austria

Locality

Urban households have higher availability of **cereals**, **vegetables**, fish and seafood, **non alcoholic beverages** and **juices** and the lowest availability of added **lipids**. In semi-urban zones the availability of **fruits** and **alcoholic beverages** are higher than in urban and rural households. The highest availability of **potatoes**, **meat**, milk products and sugar products is recorded in households in rural areas. (Figure 22 and 23)

FOOD AVAILABILITY IN AUSTRIA

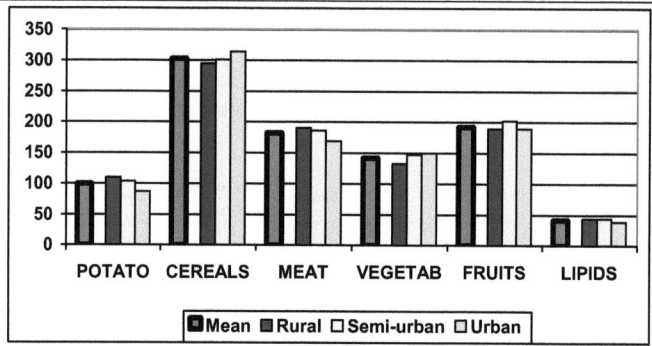

Figure 22. Average food availability, by locality (g/p/d)

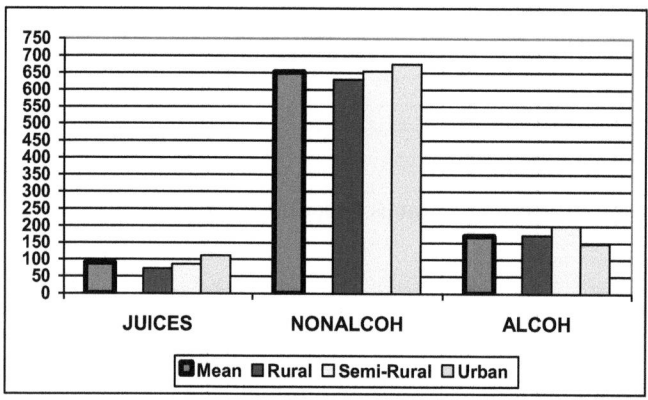

Figure 23. Average food availability, by locality (ml/p/d)

Availability of cereals and cereal products

Cookies, biscuits, wafers

Urban households show a mean availability of 21 g/p/d, whereas rural households show a daily availability (15 g/p/d).

The highest daily availability of cakes and pastries was recorded by households in urban areas (21 g/p/d), whereas the lowest mean availability was recorded by rural households (11 g/p/d).

Dark bread

Urban households show a mean availability of 59 g/p/d, whereas rural households recorded a daily availability of 72 g/p/d.

White bread

Rural and semi-urban households recorded lower daily availabilities, respectively 22 g/p/d and 23 g/p/d, than urban households (29 g/p/d).

Rice

Rural and semi-urban households show a lower daily availability of rice, respectively 12 and 14 g/p/d, whereas urban households recorded a higher mean availability of rice (18 g/p/d).

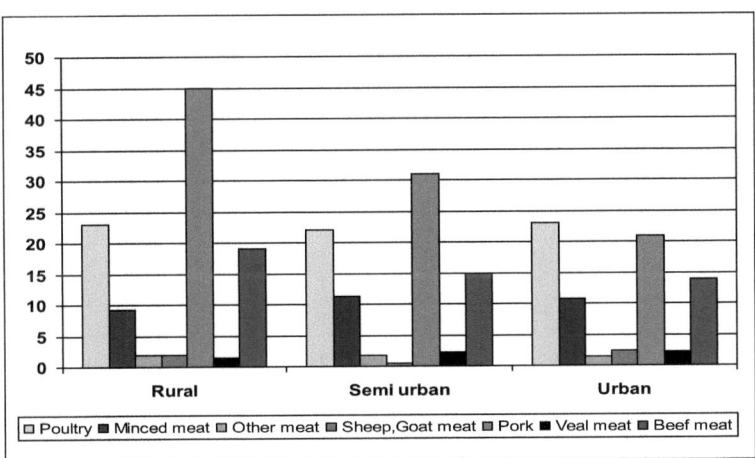

Figure 24. Meat and meat products, by locality (g/p/d)

FOOD AVAILABILITY IN AUSTRIA

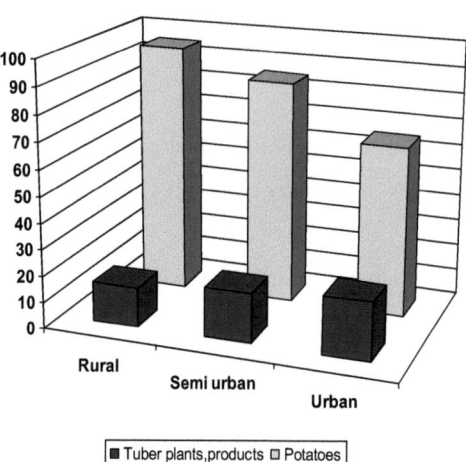

Figure 25. Potatoes and tuber plants, by locality (g/p/d)

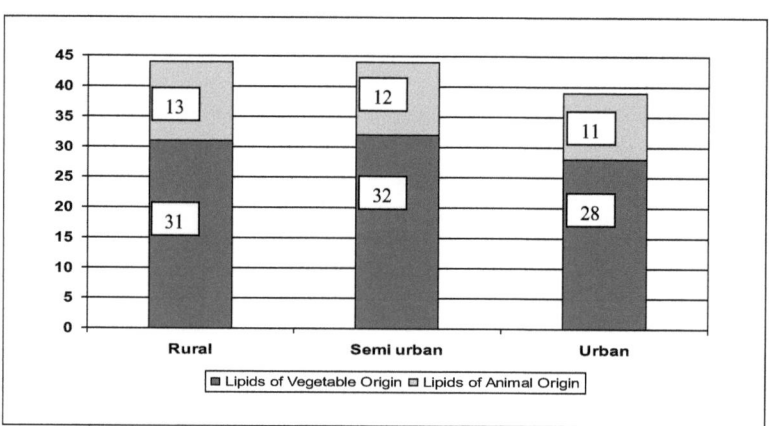

Figure 26. Lipids of animal and vegetables origin, by locality (g/p/d)

FOOD AVAILABILITY IN AUSTRIA

Availability of dairy products

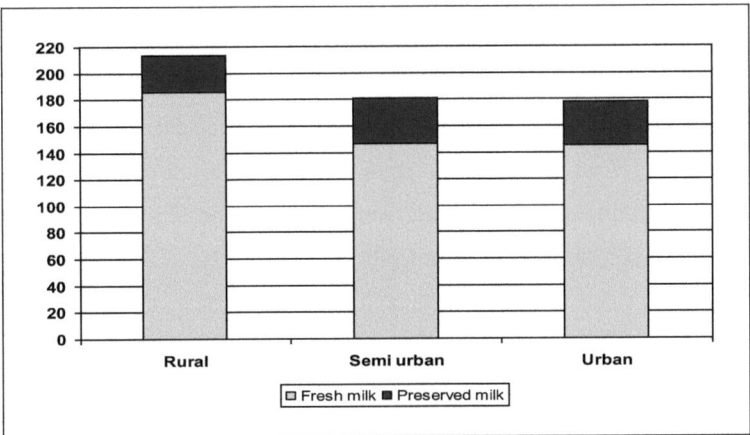

Figure 27. Fresh and preserved milk, by locality (ml/p/d)

Cheese and curd cheese

Rural and semi-urban households have lower mean cheese availabilities (respectively 17 and 18 g/p/d) than urban households (21 g/p/d).

Rural households recorded the lowest mean availability of curd cheese (9,38 g/p/d). The highest availability recorded the urban households (9,69 g/p/d).

Ice-cream

Rural and semi-urban households recorded a lower daily availability (16 ml/p/d) than urban households (22 ml/p/d).

Yoghurt

Urban households recorded an availability of 46 g/p/d of yoghurt, whereas rural households noted a daily availability of 32 g/p/d.

Availability of sugar, sweetener and sugar products

Urban households recorded the lowest value of sugar availability (23g/p/d) of all the three locality categories. Rural households noted the highest availability (46 g/p/d). Rural and semi-urban households recorded a lower mean availability of sugar products, respectively 34 and 37 g/p/d. Urban households recorded a higher availability (47 g/p/d).

The highest value of chocolate availability was recorded by urban households (29g/p/d), whereas the lowest availability was recorded by rural households (18g/p/d).

Availability of beverages and stimulants

Mineral water and soft drinks

The highest mean availability was recorded by rural households (214 g/p/d), whereas lower daily availabilities were recorded by urban and semi-urban households, respectively 206 g/p/d and 207 g/p/d.

Rural HH recorded 106 g/p/d of soft drinks; semi-urban and urban households recorded higher daily availabilities, respectively 124 and 122 g/p/d.

Coffee

Rural households recorded the lowest daily availability (15 g/p/d), whereas urban households show a higher mean availability (18 g/p/d).

Beer, non-alcoholic beer

Urban households present the lowest mean availability of beer (93 ml/p/d). The highest daily availability was recorded by semi-urban households (146 ml/p/d).

Sparkling wine, champagne, vermouth

The lowest mean availability was recorded by urban households (10 ml/p/d), whereas rural households recorded the highest daily availability (17 ml/p/d).

7.4. MEAN AVAILABILITY BY OCCUPATION

Data retrieved from the DAFNE databank, Figures - see data on p. A18 – A20

Occupation AUS_1999	1.Manual	2.Non-manual	3.Retired	4.Unemployed	5.Others
Eggs (pieces/p/d)	0,49	0,39	0,70	0,45	0,39
Potatoes (g/p/d)	90	75	156	119	89
Cereals (g/p/d)	284	284	359	313	266
Milk and milk products (g/p/d)	255	263	347	277	280
Meat and meat products (g/p/d)	167	153	253	200	145
Vegetables (g/p/d)	112	131	197	146	114
Fish and seafood (g/p/d)	6,84	8,97	12	10	8,57
Fruits (g/p/d)	156	167	278	181	155
Lipids, added (g/p/d)	41	33	62	44	35
Beverages, alcoholic (ml/p/d)	170	142	230	192	123
Beverages, non alcoholic (ml/p/d)	659	597	763	652	547
Sugar & sugar products (g/p/d)	65	62	106	64	62
Fruit & vegetable juices (ml/p/d)	87	95	81	124	97

Table 21. DAFNE main food groups, by occupation in Austria

Occupation of the household head

The retired population has the highest availability of **potatoes, cereals, meat, vegetables, fruits,** added **lipids,** fish and seafood, milk products and eggs and the lowest availability of juices in comparison with household heads of the other occupation levels. [Comment: This is a picture observed in the vast majority of the DAFNE countries. Households of retired members seem to accumulate food, with the only exception of juices]

Non manual workers have the lowest availability of added **lipids, potatoes** of all five DAFNE occupation levels. The unemployed household heads have the highest availability of juices; manual workers have the lowest availability of fish and seafood.

Unemployed households have higher daily food availability of nearly all food groups, except eggs, non alcoholic beverages, sugar and sugar products than manual and non-manual households. Manual households show higher daily food availability of these three food items than non manual households. **(Figure 28)**

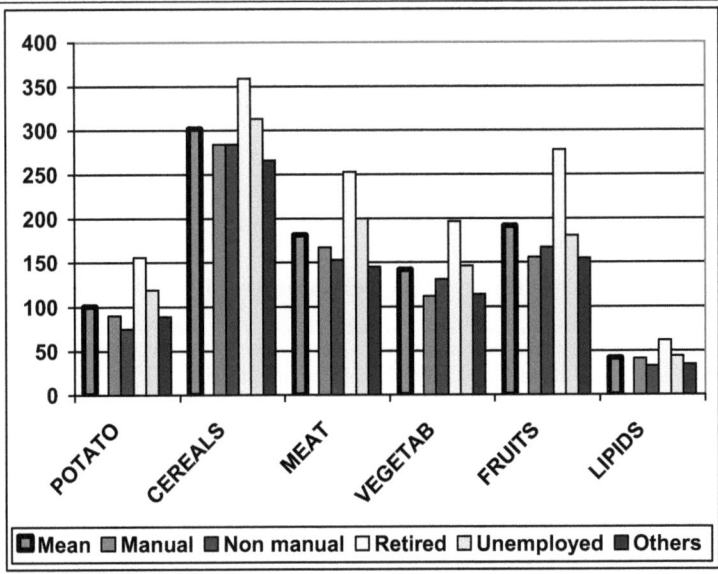

Figure 28 Average food availability, by occupation of the household head (g/p/d)

Availability of cereal products

Bread and Rolls

Retired household heads recorded the highest daily availability (119 g/p/d). The category "others" show the lowest mean availability of all occupation categories (73 g/p/d).

White bread

Unemployed household heads recorded a mean availability of 33 g/p/d which is the highest value of all categories. Whereas the category "others" recorded the lowest mean availability (21 g/p/d).

Dark bread

Retired household heads show an availability of 92 g/p/d, whereas household heads of the category "others" recorded a daily availability of 52 g/p/d.

Cookies, biscuits, wafers

The lowest mean availability of cookies was recorded by the categories "unemployed" and "manual", respectively 14 and 15 g/p/d. Retired household heads show the highest mean availability (20 g/p/d).

Rice

Households of the category "others" noted the lowest mean availability of rice (11 g/p/d). Retired persons have the highest daily availability of rice (18 g/p/d).

Flour

Unemployed household heads recorded a daily availability of 24 g/p/d which is the lowest availability of all occupation categories. The highest availability shows the category "retired" with an availability of 60 g/p/d.

Pasta

The lowest availability was recorded by household heads of the categories "manual workers" and "others" respectively 18 g/p/d and 19 g/p/d. The category "retired" noted an availability of 24 g/p/d of pasta.

Availability of milk and milk products

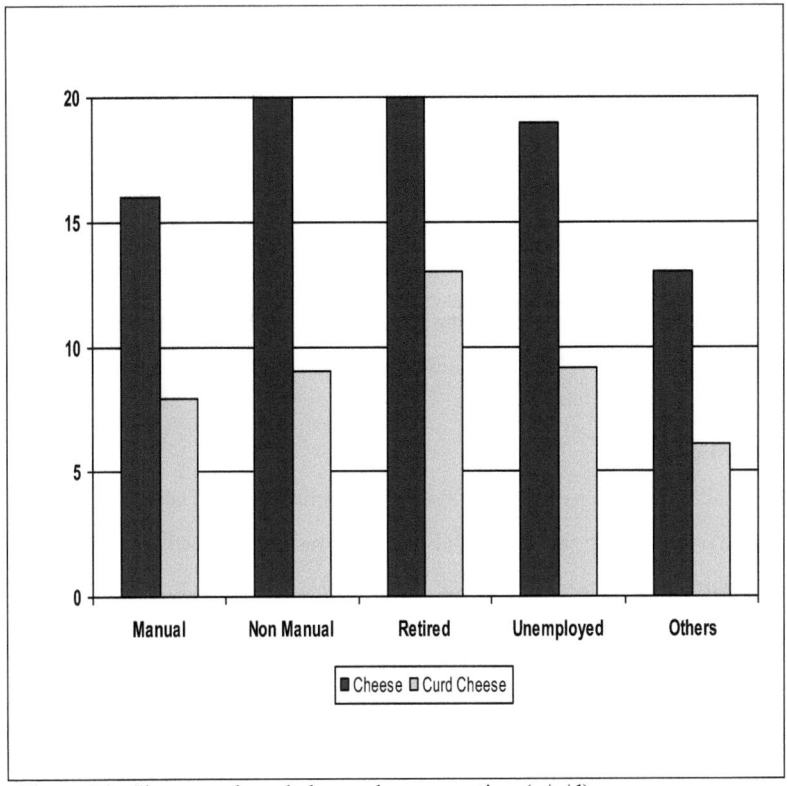

Figure 29. Cheese and curd cheese, by occupation (g/p/d)

Fresh milk

Manual workers recorded the lowest mean availability (142 g/p/d), whereas retired household heads noted the highest availability (207 g/p/d).

Preserved milk

Non manual household heads noted an availability of 24 g/p/d, whereas unemployed household heads noted an availability of 49 g/p/d.

Yoghurt

Manual workers and unemployed household heads recorded an availability of respectively 33 g/p/d, whereas retired household heads noted an availability of 41 g/p/d.

Availability of vegetables

Carrots

The category "others" shows the lowest mean availability of carrots (6,71 g/p/d), the category "retired" shows the highest daily availability (13 g/p/d).

Dried vegetables

Manual workers recorded the lowest availability of dried vegetables (0,96 g/p/d). The highest daily availability (3,66 g/p/d) "consumed" the category "unemployed".

Preserved, frozen vegetables

The lowest availability of preserved vegetables was recorded by the manual workers (20 g/p/d). The highest availability was recorded by the retired Austrians (32 g/p/d).

FOOD AVAILABILITY IN AUSTRIA

Availability of fruits

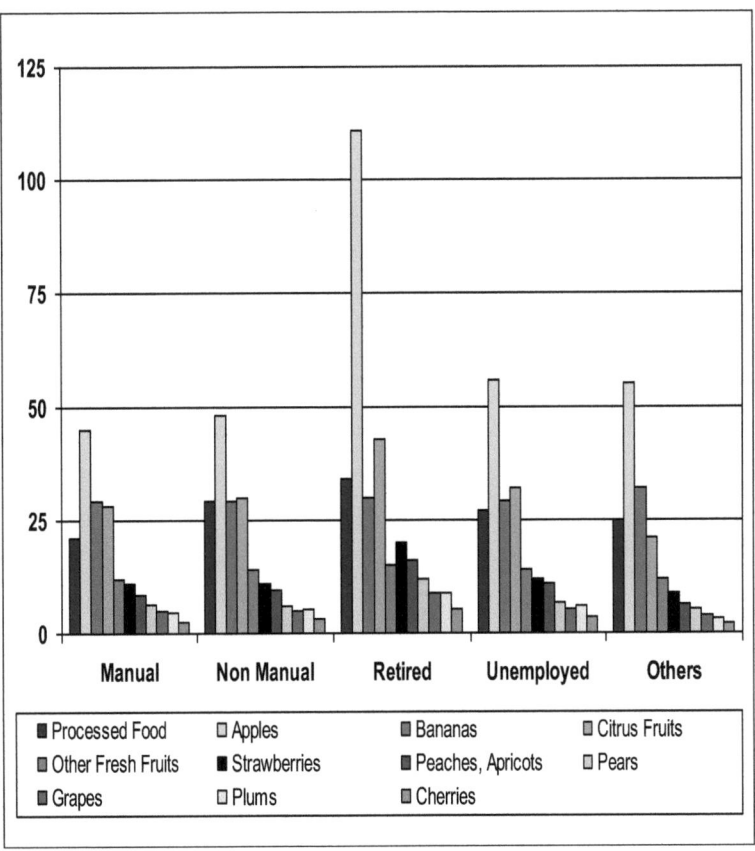

Figure 30. Average food availability, by occupation of the household head (g/p/d)

Availability of sugar and sugar products

Chocolate

Unemployed Austrian household heads recorded the lowest mean availability of chocolate (15 g/p/d). The highest availability was recorded by the retired household heads (24 g/p/d).

Jam, honey

Household heads of the category "retired" show the highest availability of jam, honey (16 g/p/d). Unemployed household heads recorded daily availability (6,68 g/p/d).

Sugar, sweetener

Household heads of the categories "non manual" and "others" recorded a daily availability of 24 g/p/d. The highest daily availability was recorded by the retired household heads (59 g/p/d).

Availability of fish and seafood

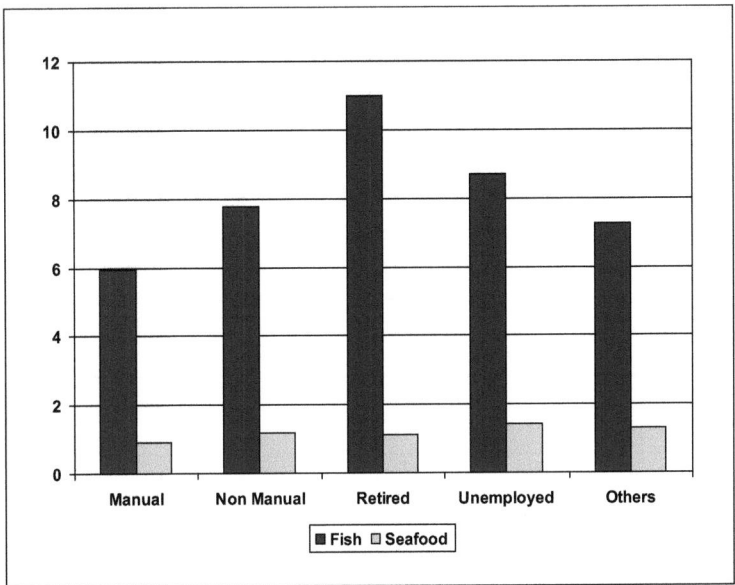

Figure 31. Fish and seafood, by occupation (g/p/d)

Availability of added lipids

Lipids of animal origin

Retired Austrian household heads recorded the highest availability of lipids of animal origin (18 g/p/d), whereas the category "others" recorded the lowest availability (8,12 g/p/d).

Butter

Household heads of the category "others" recorded the lowest availability (7,46 g/p/d), household heads of the category "retired" recorded the lowest availability (16 g/p/d) of all occupation categories.

FOOD AVAILABILITY IN AUSTRIA

Lipids of animal origin (butter excluded)

The lowest availability recorded the categories "non manual" (0,63 g/p/d) and "others" (0,66 g/p/d), whereas the highest availability was recorded by households heads of "retired".

Lipids of vegetable origin

Non manual workers have the lowest availability (23 g/p/d). Retired heads recorded the highest availability (44 g/p/d).

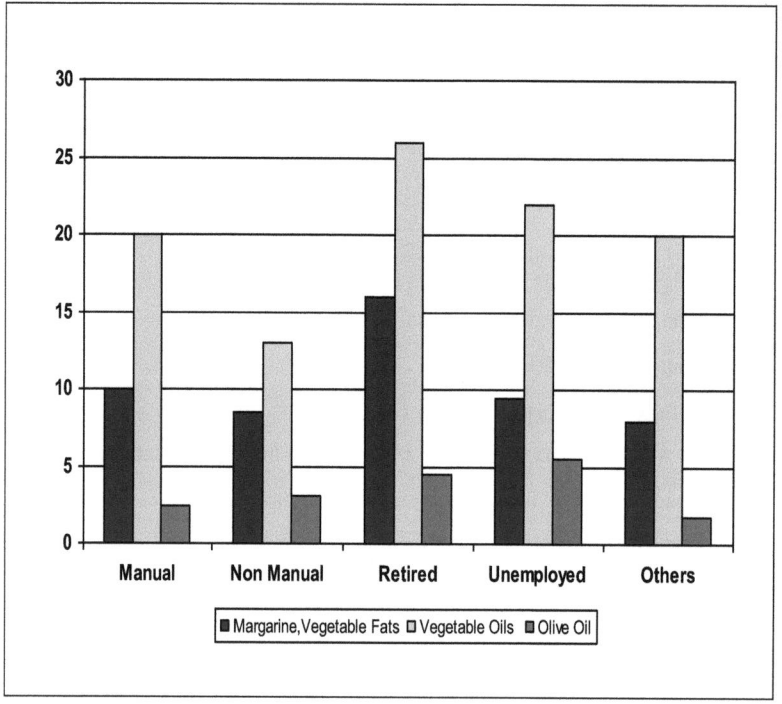

Figure 32. Margarine (g/p/d), vegetable oils and olive oil (ml/p/d), by occupation

Availability of beverages

Mineral Water

The category "others" recorded the lowest availability of mineral water (162 ml/p/d), followed by the category "unemployed" (164 ml/p/d). The highest daily availability of mineral water show the categories "manual" and "non manual" household heads (respectively 197 ml/p/d).

Soft drinks

Retired household heads recorded 92 ml/p/d of soft drinks. The category "others" recorded the highest availability (138 ml/p/d).

Beer

Non manual workers show the lowest availability of beer (92 ml/p/d), whereas retired and unemployed household heads show the highest availability (154 ml/p/d).

Schnapps, spirits, liqueurs

Manual workers recorded an availability of 2,68 ml/p/d, whereas retired household head recorded a higher availability (7,63 ml/p/d).

Red wine, Rose

The lowest availability was recorded by the category "others" (7,09 ml/p/d). 20 ml/p/d is the highest daily availability of all occupational categories which was recorded by the retired household heads.

White wine

Unemployed household heads recorded a daily availability of 7,61 ml/p/d. The highest mean availability was shown by the retired household heads (27 ml/p/d).

7.5. MEAN AVAILABILITY BY HOUSEHOLD COMPOSITION

Data retrieved from the DAFNE databank, Figures - see data on p. A21 – A23

Household composition _AUS_1999	1. Adult household- single	2. Adult household- 2 members	3. Adult+ children (lone parents)	4. Adults+ children
Eggs (pieces/p/d)	0,57	0,52	0,43	0,39
Potatoes (g/p/d)	86	103	93	76
Cereals (g/p/d)	372	310	314	264
Milk and milk products (g/p/d)	342	277	288	249
Meat and meat products(g/p/d)	202	216	137	139
Vegetables (g/p/d)	161	177	115	107
Fish and seafood (g/p/d)	8,72	13	10	7,21
Fruits (g/p/d)	252	242	172	143
Lipids, added (g/p/d)	44	47	39	31
Beverages, alcoholic (ml/p/d)	244	218	87	120
Beverages, non alcoholic (ml/p/d)	950	757	608	530
Sugar & sugar products (g/p/d)	85	79	65	58
Fruit & vegetable juices (ml/p/d)	127	96	110	91

Table 22a. DAFNE main food groups, by household composition in Austria

Household composition _AUS_1999	5. Adult+ elderly	6. Adult+elderly+children	7. Elderly – single	8. Elderly-2 members	9. Others
Eggs (pieces/p/d)	0,69	0,49	0,80	0,60	0,50
Potatoes (g/p/d)	156	82	167	143	118
Cereals (g/p/d)	368	291	383	357	282
Milk and milk products (g/p/d)	350	286	393	321	264
Meat and meat products (g/p/d)	251	154	221	213	222
Vegetables (g/p/d)	181	113	216	196	150
Fish and seafood (g/p/d)	11	5,82	12	14	8,73
Fruits (g/p/d)	226	161	329	261	174
Lipids, added (g/p/d)	60	44	77	62	40
Beverages, alcoholic (ml/p/d)	255	170	184	232	173
Beverages, non alcoholic (ml/p/d)	735	555	818	673	665
Sugar & sugar products (g/p/d)	100	73	134	93	68
Fruit & vegetable juices (ml/p/d)	74	50	90	69	86

Table 22b. DAFNE main food groups, by household composition in Austria

Household composition

The food and beverage availability of the household's composition related to number and age of the household members were also analysed.

Single adult households have higher daily food availability of eggs, cereals, milk products, fruits, alcoholic beverages, non alcoholic beverages, sugar and sugar products,
fruit and vegetable juices than two member adult households. Two member adult households have higher availability of potatoes, meat, fish and seafood, vegetables, added lipids than single adults.

Lone parent households show higher availability of eggs, potatoes, cereals, milk and milk products, vegetables, fish and seafood, fruits, added lipids, non alcoholic beverages, sugar and sugar products, fruit and vegetable juices and lower availability of meat, alcoholic beverages than two adults and children households.

Single elderly households have higher availability of eggs, potatoes, cereals, milk and milk products, meat, vegetables, fruits, added lipids, non alcoholic beverages, sugar and sugar products, fruit and vegetable juices and lower availability of fish and seafood and alcoholic beverages than two elderly households.

Single elderly households have higher availability of eggs, potatoes, cereals, milk products, vegetables, fruits, added lipids, sugar and sugar products than the other household composition categories. Two elderly households have higher availability of fish and seafood than two adult households. The highest availability of non alcoholic beverages and juices is recorded in single adult households.

FOOD AVAILABILITY IN AUSTRIA

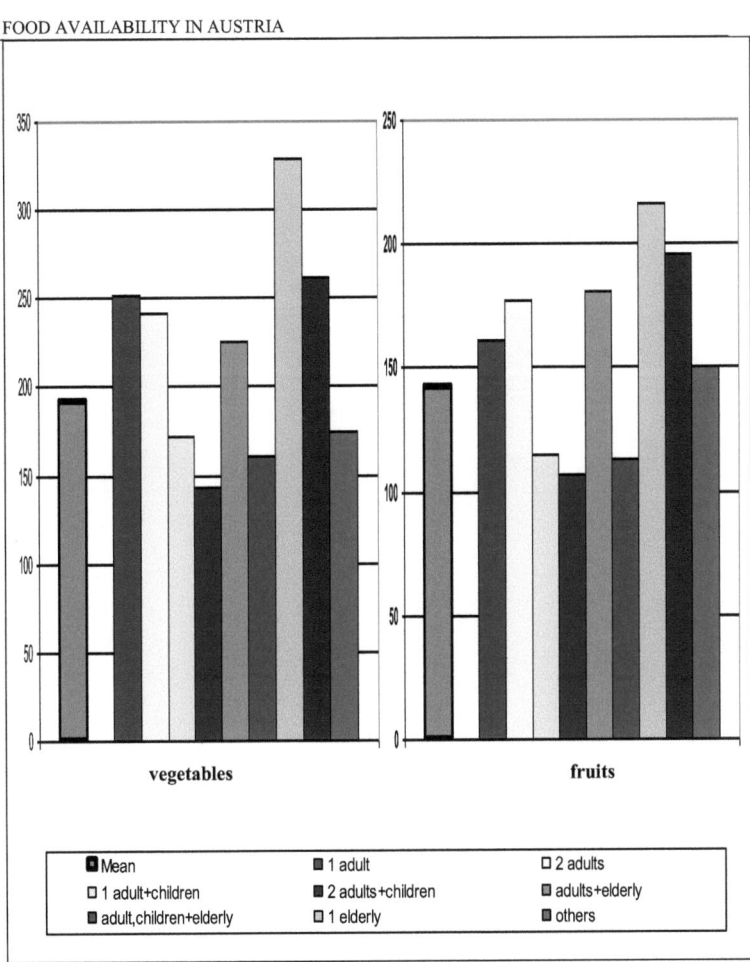

Figure 33. Average food availability, by household composition (g/p/d)

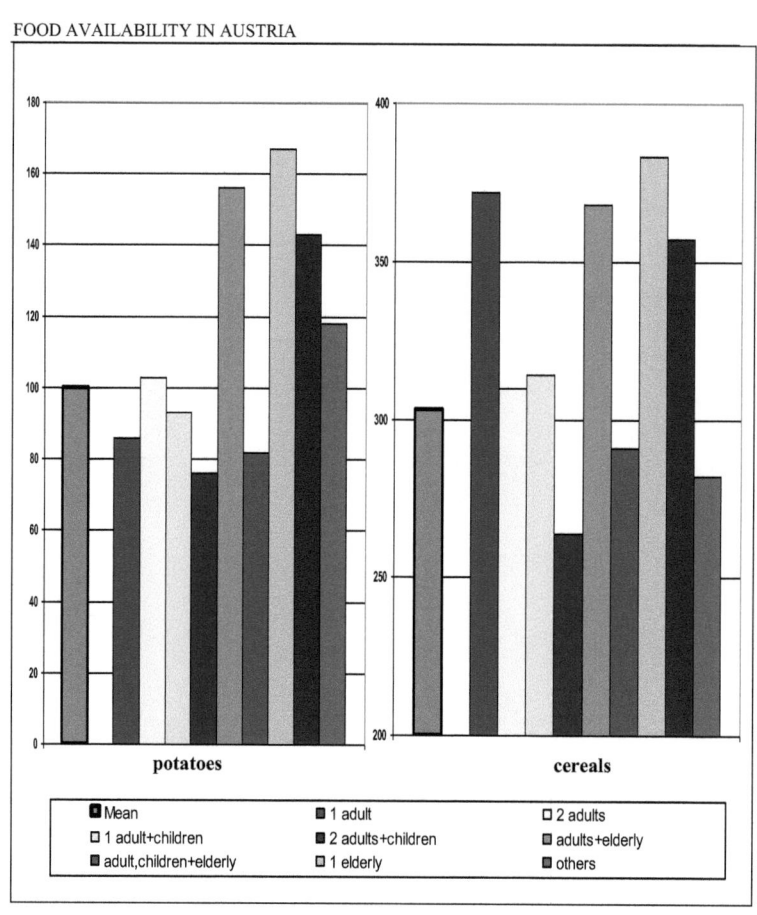

Figure 34. Average food availability, by household composition (g/p/d)

FOOD AVAILABILITY IN AUSTRIA

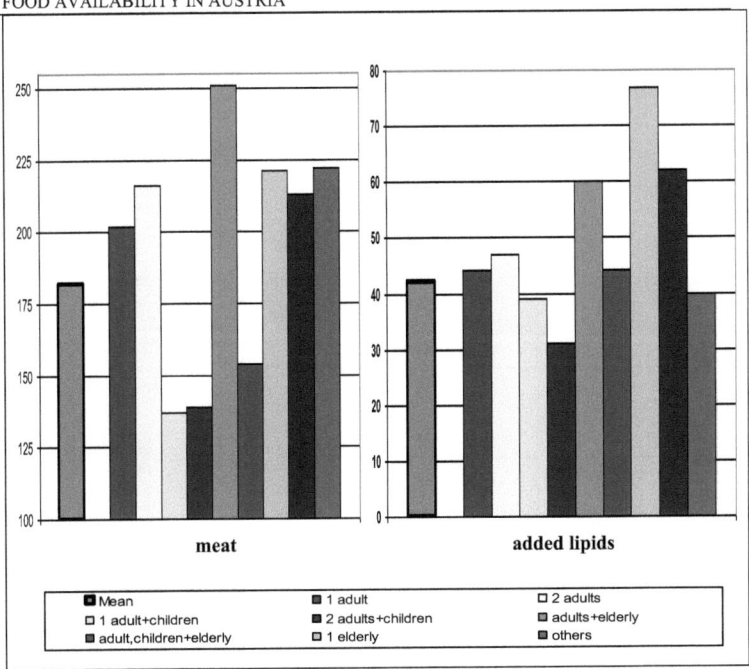

Figure 35. Average food availability, by household composition (g/p/d)

Availability of meat and meat products

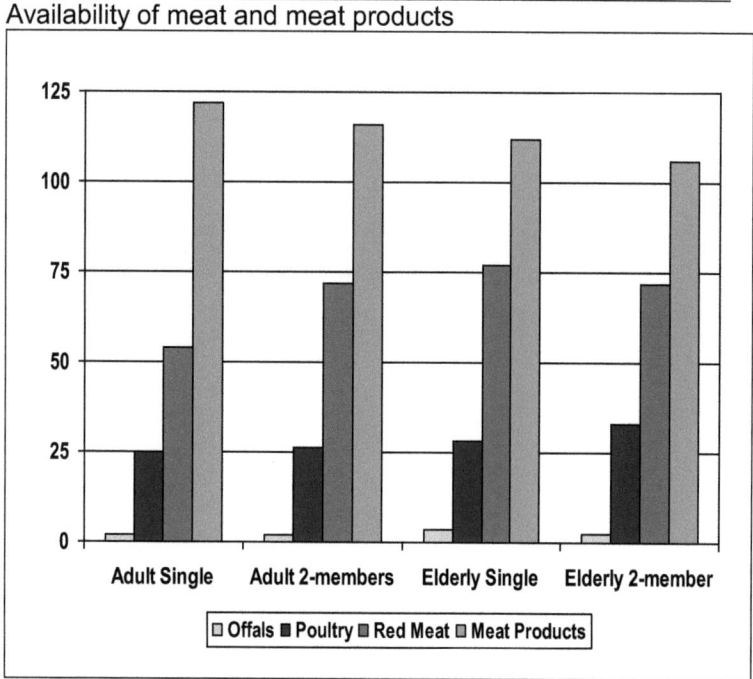

Figure 36. Red meat, poultry, meat products and offals, by household composition (g/p/d)

Sausages

Single households of the categories "adult" (37 g/p/d) and "elderly" (38 g/p/d) show lower mean availability values than 2-member households of the categories "adult" (40 g/p/d) and "elderly" (41 g/p/d).

Smoked products

Adult single households recorded the lowest daily availability (12 g/p/d), whereas elderly single HH recorded the highest mean availability (19 g/p/d). Adult 2-member households show an availability of 13 g/p/d, elderly 2-member HH present 16 g/p/d.

Availability of sugar products

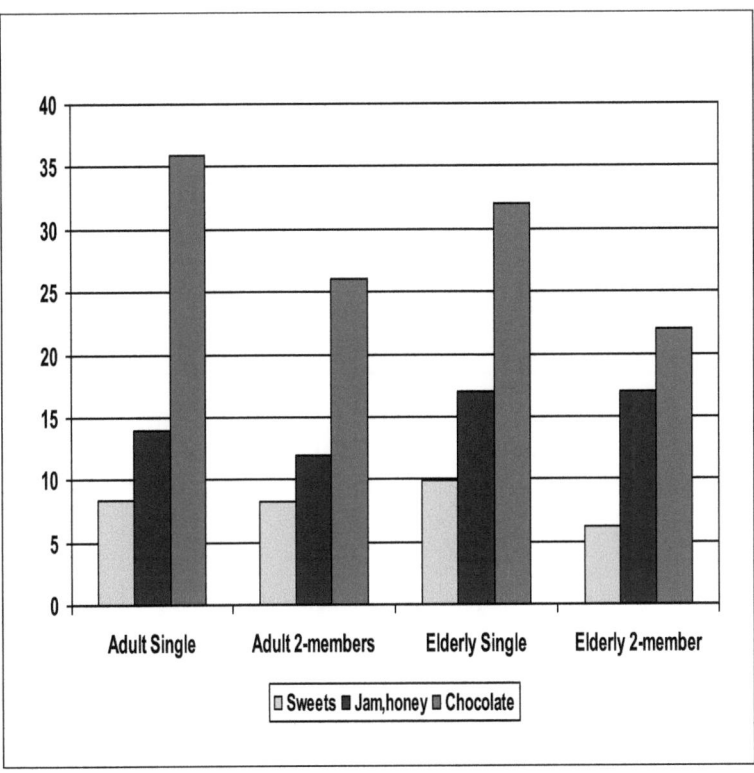

Figure 37. Chocolate, sweets, jam and honey, by household composition (g/p/d)

Availability of vegetables

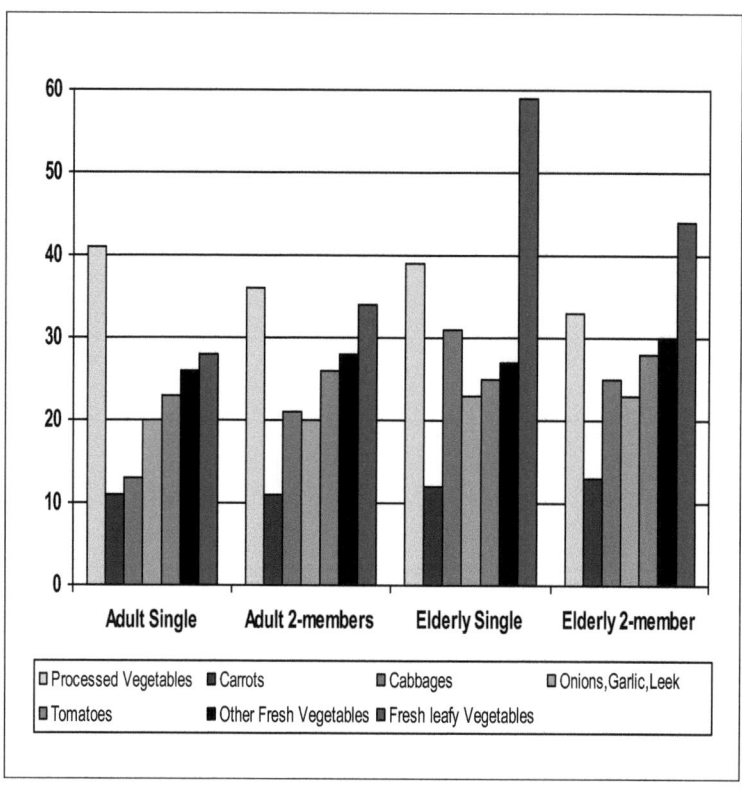

Figure 38. Average food availability, by household composition (g/p/d)

FOOD AVAILABILITY IN AUSTRIA
Availability of cereals and cereal products

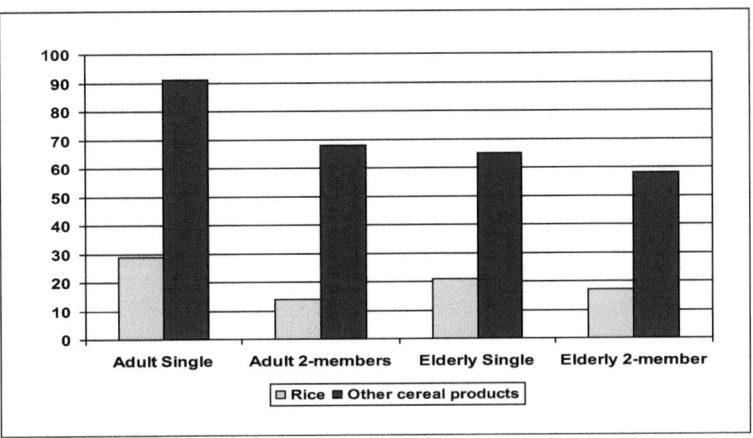

Figure 39. Rice and other cereal products, by household composition (g/p/d)

Cakes and pastries

Single households present higher daily availability values of cakes and pastries than 2-member households. Adult single households show a lower mean availability (23 g/p/d) than elderly single households (24 g/p/d). Adult 2-member households recorded present a lower daily availability (18 g/p/d) than 2-members (19 g/p/d).

Dark bread

Single households recorded higher daily availability values than 2-member households. Adult single households present a daily availability of 81 g/p/d, whereas adult 2-member households show an availability of 68 g/p/d. Elderly single households recorded an availability of 101 g/p/d, elderly 2-member households recorded a mean availability of 95 g/p/d.

White bread
Single households recorded higher daily availability values than 2-member households.
Adult single and elderly single households recorded a daily availability of 32 g/p/d. Adult 2-member present an availability of 30 g/p/d, elderly 2-member households show an availability 27 g/p/d.d

Pasta
Adult single households show the highest mean availability of pasta (34 g/p/d), whereas adult 2-member households recorded the lowest mean availability (19 g/p/d).
Elderly single households present a higher availability of pasta (26 g/p/d) than elderly 2-member households (25 g/p/d).

Availability of dairy products
Cheese
Elderly households present lower daily availability values than adult households. Elderly 2-member households recorded a lower availability (19 g/p/d) than elderly single households (22 g/p/d). Adult single households record a higher availability of cheese (27 g/p/d) than adult 2-member households (23 g/p/d).

Ice-cream
Elderly 2-member households recorded a daily availability of 9,96 ml/p/d, whereas elderly single households recorded a mean availability of 15 ml/p/d. Adult 2-member household represent a lower daily availability of ice-cream (19 ml/p/d) than adult single households (20 ml/p/d).

Yoghurt

Elderly 2-member households present the lowest daily availability of yoghurt (36 g/p/d), elderly single households recorded an availability of 49 g/p/d. Adult single households present the highest mean availability (61 g/p/d), adult 2-members recorded an availability of 45 g/p/d.

Availability of beverages and stimulants

Mineral Water

Single households recorded higher mean availability values than 2-member households. Whereby elderly single households present a higher daily availability (286 ml/p/d) than adult single households (267 ml/p/d). Elderly 2-member households present a lower mean availability (241 ml/p/d) than adult 2-member households (255 ml/p/d).

Cocoa

Elderly Single households have a higher daily availability of cocoa (2,92 g/p/d) than elderly 2-member households (1,69 g/p/d. Adult single households show a lower mean availability (1,2 g/p/d) than adult 2-member households (2,12 g/p/d).

Sparkling wine, champagne, vermouth

Elderly households with two members present an availability of 9,11 ml/p/d, whereas elderly single households present a mean availability of 20 ml/p/d.

Adult single households recorded a lower availability of 14 ml/p/d; adult 2-member households recorded a mean availability of 17 ml/p/d.

Availability of added lipids

Lipids of animal origin

Adult single households present the lowest daily availability 12 g/p/d. The highest availability was recorded by elderly single households (24 g/p/d). Adult 2-member households present an availability of 13 g/p/d, whereas elderly households with two members show an availability of 18 g/p/d.

Animal fat

The highest mean availability was recorded by elderly 2-member households 2,87 g/p/d. The lowest mean availability was recorded by adult single households (0,62 g/p/d). Adult 2-member households recorded an availability of 1,46 g/p/d. Elderly single households present a mean availability of animal fat 2,35 g/p/d.

Butter

Elderly single households present the highest availability of butter (21 g/p/d). Adult single households show the lowest mean availability (11 g/p/d). Adult 2-member households recorded a daily availability of 12 g/p/d, elderly 2-member households show an availability of 15 g/p/d.

Lipids of vegetable origin

The highest mean availability was recorded by elderly single households (54 g/p/d).
The lowest availability was recorded by adult single households 32 g/p/d. Adult households with two members present a daily availability of 33 g/p/d. Elderly 2-members show a mean availability of 45 g/p/d.

Olive oil

The lowest daily availability was recorded by adult households with two members (4,28 g/p/d), whereas the highest availability was recorded by adult single households (6,75 g/p/d). Elderly single households present a daily availability of 5,76 g/p/d, whereas elderly 2-member households show an availability of 5,93 g/p/d.

Other salad, cooking oil

The lowest mean availability values were recorded by adult single households and adult 2-member (19 g/p/d). The highest availability value was recorded by elderly single households (32 g/p/d). Elderly 2-member households present a daily availability of 23 g/p/d.

8 RESULTS – FOOD AVAILABILITY IN EUROPE

One of the most obvious ways in which the people of Europe differ is in what they eat (McKee M. and Ryan J., 2003, p.2).

Since 1989 Europe has experienced dramatic changes in its political and economic conditions. The cost of food as well as the choice of food items has increased, and there have been profound changes in personal incomes, their distribution and the socio-economic environment in which food choices are made (Szponar L. et al., 2001, p.1183).

The European Union is an excellent area to expand their food premises, but they need good knowledge about food consumption in different countries. In different countries exist similarities and dissimilarities. Some of them are related to classical economic factors, others to actual lifestyles, and many others to socio-demographic characteristics.

Consumption patterns differ among European countries, although there are common trends, and constitute a good example of how to find a compromise between global and local trends (Gracia A. and Albisu L.M., 2001, p.486).

Total food consumption is determined by the consumption of individuals and the number of inhabitants of a certain region and their demographic structure. Demographic trends influence food marketing by affecting the number of mouths to feed what people eat and how people buy their food (Ratinger T. and Šlaisova J., 2001, p.175-176).

FOOD AVAILABILITY IN EUROPE

The European Union is an area characterized by a high development status under an internal integration process and, at the same time, external internationalization opening.

Consumers are becoming wealthier and they are affluent enough so that they demand food not only for nourishment reasons but also enjoyment, preference to be very important in building final food choices. Economic, social, and demographic characteristics and food consumer choice and behaviour subscribe European food consumption (Gracia A. and Albisu L.M., 2001, p.469).

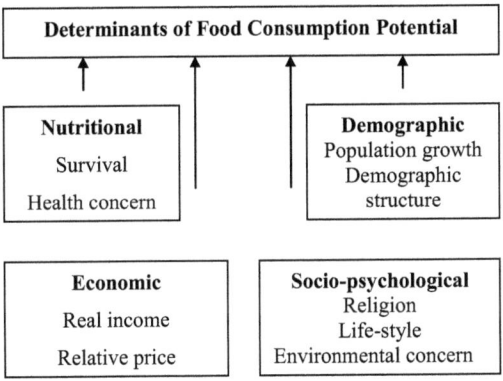

Figure 40. Determinants of food consumption potential (Ratinger T. and Šlaisova J., 2001, p.176)

Food consumption in the European Union countries can be summarized in four major trends:

(1) A decrease in the proportion of expenditure allocated to food already reaching very low levels (Gracia A. and Albisu L.M., 2001, p.470).
In recent decades there have been significant changes in the foods that we eat, the way that they are eaten and how much we spend on food. The proportion of household income spent on food has dropped from 50% to 25% over the last 40 years across Europe as a whole, and there has been a huge increase in the proportion of food consumed outside the home (Société française, 2000, p.5-6).

(2) A maximum level in total food consumption, in quantity terms.
The second trend is the result of a situation that occurs in wealthy countries where quantity is surpassed by quality concerns; people want to eat better as their daily intake requirements diminish.

(3) A shift in the food consumption structure.
The third trend is not as homogeneous, and it differs from country to country, according to many distinctive aspects, but also based on their cultural and historical evolution.

(4) An increase in the proportion of food consumed away from home.
The fourth trend is common for all countries, but intensity varies among countries and labour circumstances.

Previous changes in consumer characteristics (rising income, aging population, smaller household, women labour participation, etc.) have caused European consumers to demand more added-value food products. In particular, they demand higher quality and more diversify products.

Consumers acting in this environment are facing important changes. They are exposed to an increasing number and more diversified quality food products.

In the early 1960s, the diet in European Mediterranean countries (Greece, Italy, Portugal, and Spain) was considered healthy and representative of the Mediterranean diet. However, the trend in food consumption in these countries has moved away from such a diet. Furthermore, some northern countries are changing their food demand towards healthy characteristics.

The proportion of working women is increasing, and it accounts for over 40% in most countries except Luxembourg, Ireland, and the Mediterranean countries Greece, Italy and Spain. Therefore, household income levels have increased, and the amount of time available for cooking has been reduced. The main consequence is the increasing use of convenience food, ready-to-eat meals, and the increasing number of meals consumed away from home (at the workplace, children at school). Despite the large number of women involved in professional activities, they are still largely responsible for family nutrition, and they are the main family meal planners (Gracia A. and Albisu L.M., 2001, p.470-478).

The average household size in the EU is diminishing, although the total number of households is increasing. Therefore, the number of single-person households is increasing, and it accounts for 11% in Spain, 34% of all private households in Austria (Statistik Austria, 2006) up to as much as 41% in Sweden in 1995. Those households can be split into two categories: pensioners who live alone, or single young people. The latter usually eat more often away from home or at workplace, and they buy more often ready-to-eat meals and try new products.

Older consumers are more conservative and prefer the food products they used to eat in the past. They rarely try either new food products or ready-to-eat meals, and they infrequently eat away from home. Their per capita income is relatively high, and they allocate an important part of their income to buy food products, although they eat less quantity than younger people do (Gracia A. and Albisu L.M., 2001, p.476).

8.1. MEAN DAILY FOOD AVAILABILITY AMONG ELDERLY EUROPEAN LIVING ALONE

Figure 41 presents the mean daily availability of eggs, milk and milk products, vegetables and fruits among male and female European aged 65 years or more, who live alone. Females recorded higher food availability for the four foods under study. Milk and milk products are widely consumed by elderly Europeans, who further reported higher fruit than vegetable availability. Given current recommendations of WHO of at least 400 g of vegetables and fruits per day, the HBS-derived daily fruit and vegetable availability for single, elderly European seems adequate. Elderly individuals, particularly women living alone, have often been reported to overpurchase during the survey period (Trichopoulou A. et al., 2005, p.72).

FOOD AVAILABILITY IN EUROPE

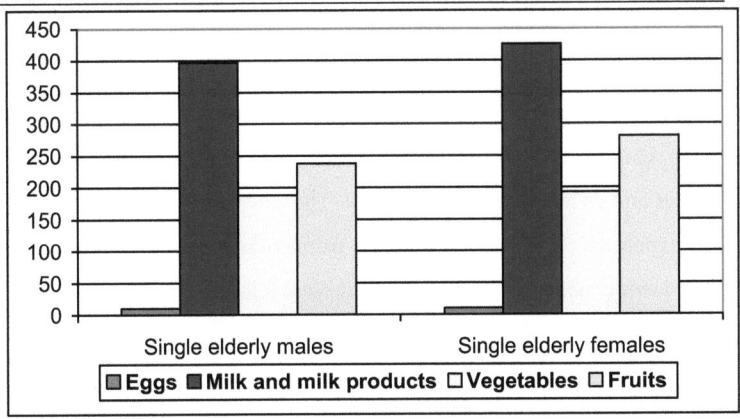

Figure 41. Mean daily individual availability of selected foods among elderly Europeans living alone (g/p/d)[1] (source dafne databank)

8.2. CULTURAL AND SOCIAL FACTORS

Health is also influenced by economic, social, and cultural conditions in a society. Thus, the interaction between diet, physical activity and health may reflect changes in the political and economic system of a country (Zunft H.-J.et al., 1999, p.437). The availability of food locally can influence food culture (Rowely R., 2006, p.2).

Cultural factors, including religious or moral beliefs or peer group expectations, clearly have an impact on food choices and dietary patterns.

---

[1] NB. Values are calculated as weighted arithmetic means of the mean availability values of ten European countries (Austria, Belgium, Finland, Germany, Greece, Ireland, Italy, Norway, Portugal an the UK)

Traditional eating patterns and attitudes to food vary considerably between and within European states.

Price and income are clearly important influences over food although the extent to which price is an influence varies from to one European country to another and different groups in society. The way in which low income groups experience "food poverty" varies from country to country. This is because a wide range of factors, including food retailing patterns, public transport networks, welfare support for low income groups and the relative prices of different foods (Société française, 2000, p.14-15).

8.3. EUROPEAN FOOD AVAILABILITY

There are disparities in dietary habits between and within European populations, resulting mostly from their different cultural norms. The dietary choices of the population subgroups may also be related to socio-economic differences, which have been reported to be implicated in the clustered patterns of disease observed among the less privileged of the European societies. The inadequate diet of the low socio-demographic groups, in combination with other poor lifestyle choices (e.g. smoking, limited physical activity), is listed among the risk factors that may partially explain the observed outcome (Trichopoulou A. et al., 2002b, p.554). The diet of the lower socioeconomic groups provides cheap energy from foods such as meat products, full cream milk, fats, sugars, preserves, potatoes, and cereals but has little intake of vegetables, fruit, and whole-wheat bread. This type of diet is lower in essential nutrients such as calcium, iron, magnesium, folate, and vitamin C than that of the higher socioeconomic groups (James W.Ph. et al., 1997, p.1545).

Socio-economic differences in eating practises are often studied in terms of the level of education achieved. Education has been reported to be the strongest and most consistent indicator in assessing socio-economic differentials. Education expresses not only the individual's attainment and years of schooling, but it might also reflect occupation, income and, even more importantly when it comes to healthy dietary practise, the way an individual perceives and applies current nutritional information.

The food purchasing capacity of the household, often expressed as proportion of the household's expenses that refer to food purchases, has been shown to indicate low socio-economic class or small income and as a measure of food security within households, since those households with a high proportion of food expenses are more vulnerable when an unexpected event (e.g. job loss, natural disaster) limits the financial capacity (Trichopoulou A. et al., 2002, p.554).

People with higher socioeconomic status often report higher consumption of vegetables and fruit than people with lower socioeconomic status. There are indications that dietary antioxidants, fibre and other components of vegetables and fruit play a role in the prevention of cardiovascular diseases and cancer. Socioeconomic differences in the dietary intake of antioxidants such as vitamin C and β-carotene have been partly proposed to explain socioeconomic differences in cardiovascular diseases (Lindström M. et al., 2001, p.51).

FOOD AVAILABILITY IN EUROPE

8.3.1. Food Availability in North, Central and South Europe

Finland, Austria, Greece
Data retrieved from the DAFNE databank

In the following chapters the daily food availability results of DAFNE Austria are compared with representatives of North-, and South-Europe (8.3.1.), with the mean DAFNE food availability (8.3.2.) and with the results of the other DAFNE members (8.3.3.).

Concerning the north-south differentiation in Finnish, Austrian and Greek food habits, the life expectancy of these representatives of North-, Central- and South-Europe were observed in chapter Discussion (see p. 130 – 131).

In the following pages graphs were selected to illustrate disparities in the availability of DAFNE food groups in Finland, Austria and Greece which should represent North, Central and South Europe.

Bearing in mind that HBS values on food and beverages are only related to household consumption, a North – South differentiation could be observed.

The once obvious differences between urban and rural population are diminishing; in part because of considerations of retail and food supply, but also because individual level rather than area level affects food consumption; influences on deprivation are relatively more important in contemporary Europe outside the cities.

FOOD AVAILABILITY IN EUROPE

It is clear that the North – South gradient in diet persists across Europe, despite increasing congruence, with high rates of consumption of meat and meat products and lower rates of fruit and vegetable intake in the northern countries (Kelleher C. et al., 2002, p. 529-531).

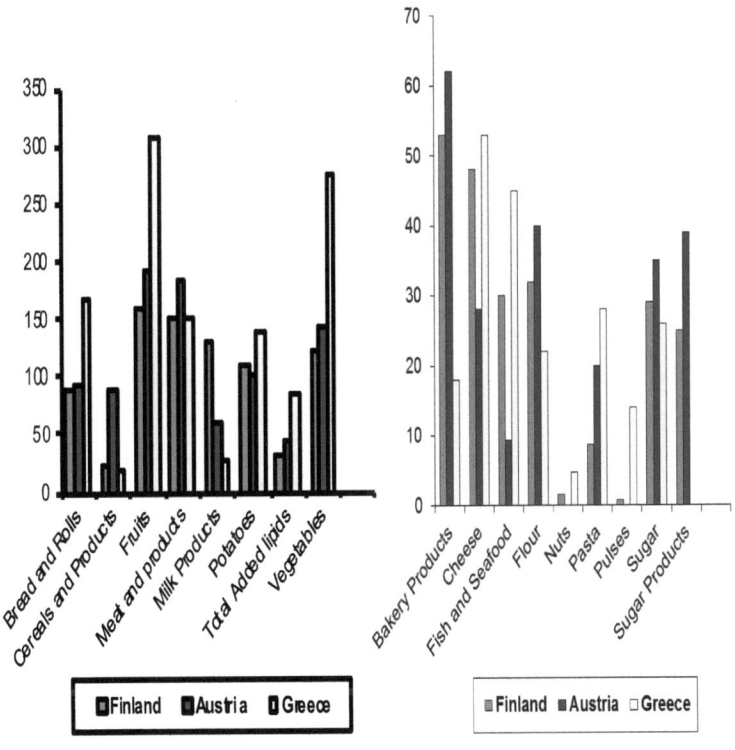

Figure 42. DAFNE main food groups in North, Central and South Europe (g/p/d)

Potatoes

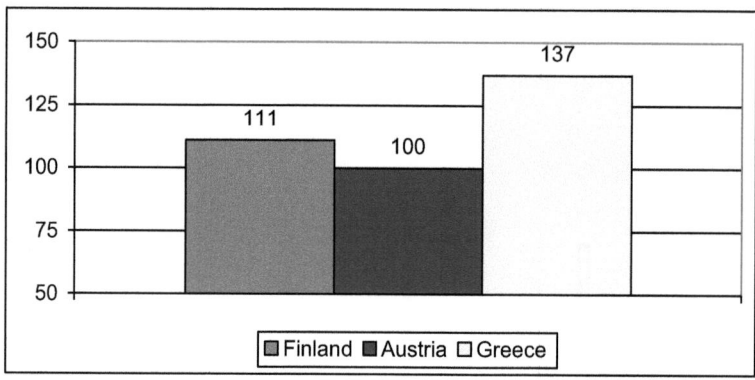

Figure 43. Daily availability of potatoes – mean (g/p/d)

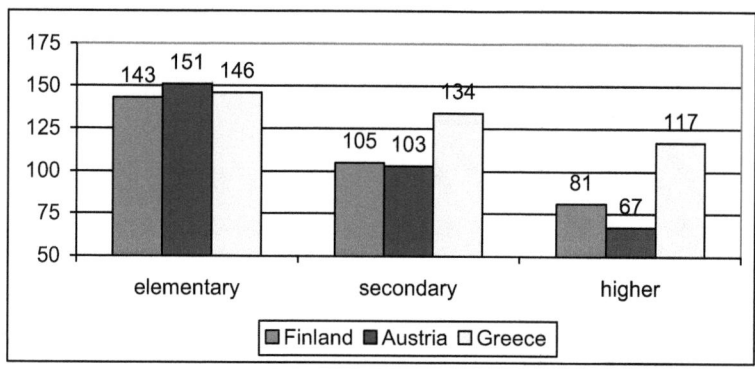

Figure 44. Daily availability of potatoes by education (g/p/d)

Greece presents the highest availability of potatoes (137 g/p/d). Austria and Finland recorded availability values of respectively 100 g/p/d and 111 g/p/d.

Finnish rural, semi-urban and urban households recorded higher daily availabilities than Austrian households. Greek, Finnish and Austrian rural households (respectively 156, 155 and 110 g/p/d) recorded higher mean availability values than the other locality categories (urban HH respectively 131, 90 and 87 g/p/d).

The highest daily availability was recorded by Austrian elementary educated households (151 g/p/d). The lowest value was recorded by Austrian higher educated households (67 g/p/d).

Greek households of the categories "manual", "non manual", "unemployed", "others" recorded the highest mean availability values of their respective category.

Austrian adult single households recorded the lowest mean availability (86 g/p/d) in comparison to the other household composition categories in Finland, Austria and Greece. Finnish elderly households with two members recorded the highest daily availability (174 g/p/d).

Meat and meat products

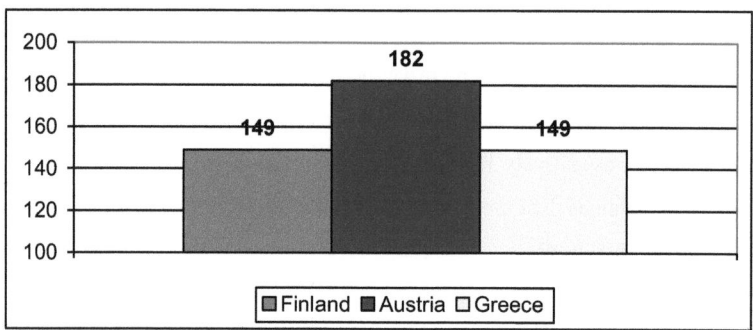

Figure 45. Daily availability of meat and meat products (g/p/d)

Austria households recorded a higher mean availability (182 g/p/d) than Finnish and Greek (respectively 149 g/p/d).

Finnish higher educated households show the lowest daily availability value (133 g/p/d).

Austrian elementary households recorded the highest mean availability (197 g/p/d).

Austrian rural, semi-urban and urban households present higher mean availability values than Finnish households. Austrian rural households recorded the highest availability (190 g/p/d), whereas Finnish urban households recorded the lowest availability (141 g/p/d).

Finnish households of the category "others" present the lowest daily availability (95 g/p/d), whereas Austrian retired households show the highest availability (253 g/p/d).

Finnish households of the household composition categories show the lowest mean availability values, whereas Austrian households recorded the highest availability. Austrian elderly single households show the highest availability (221 g/p/d); Finnish elderly households with two members recorded the lowest value (149 g/p/d).

Fruits

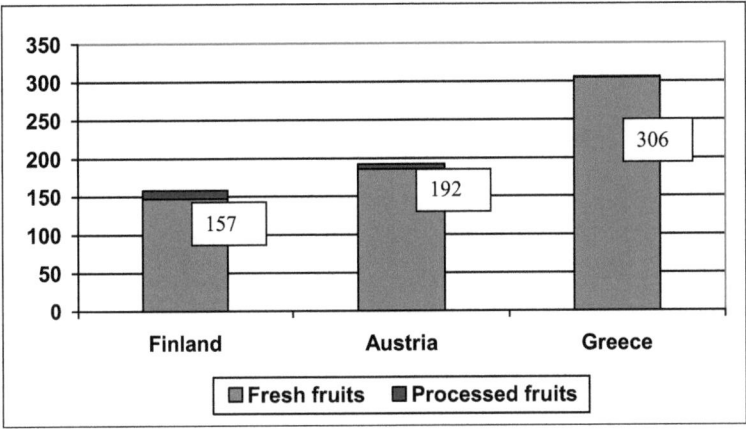

Figure 46. Daily availability of fruits – mean (g/p/d)

Greece recorded the highest mean availability of fruits (306 g/p/d), Finland present the lowest availability (157 g/p/d).

Greek households of the three education categories show the highest daily availability values, whereas Finnish households present the lowest mean availability values. Greek higher educated households recorded the highest value (357 g/p/d), Finnish secondary educated household recorded the lowest value (145 g/p/d).

Finnish households of all occupation categories recorded lower daily availability values than Greece and Austria. Greece presents higher values. The highest availability value was recorded by Greek retired households (360 g/p/d), whereas Finnish households of the category "others" recorded the lowest daily availability (106 g/p/d).

Finnish adult single households show the lowest mean availability (164 g/p/d), whereas Greek adult single households recorded the highest availability (458 g/p/d) of all occupation categories.

Vegetables

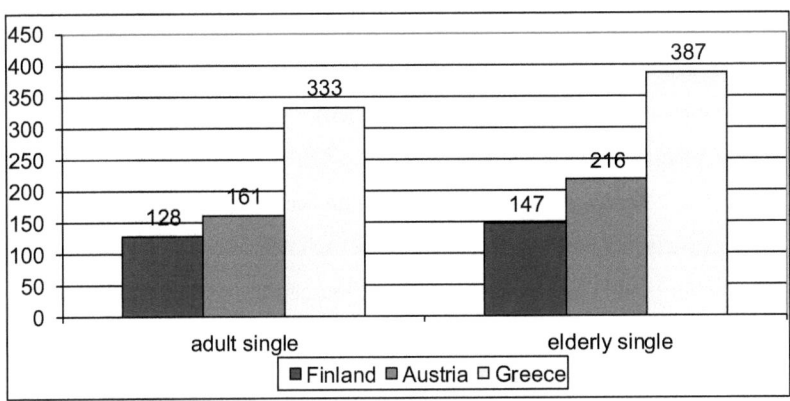

Figure 47. Daily availability of vegetables, by household composition (g/p/d)

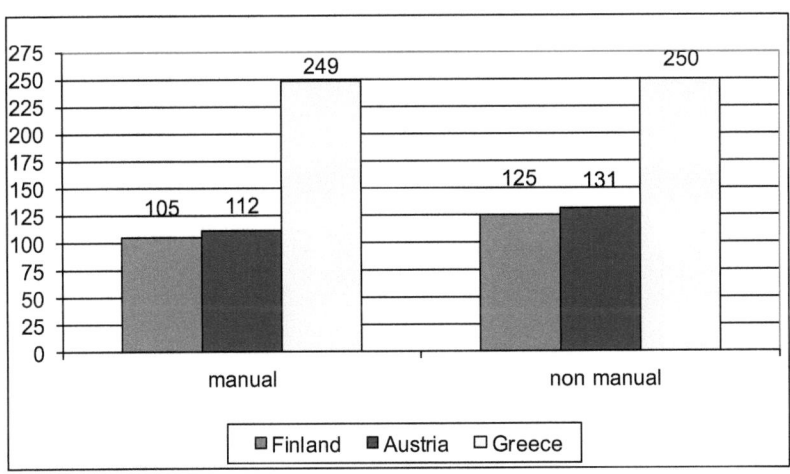

Figure 48. Daily availability of vegetables, by occupation (g/p/d)

Greek households show the highest average daily availability of vegetables (274 g/p/d), whereas Finland present the mean lowest availability (122 g/p/d).

Finnish adult single households present the lowest daily availability of vegetables (128 g/p/d), whereas Greek elderly single households show the highest availability (387 g/p/d).

All Greek occupation categories present higher daily availability values than Austria and Finland. Greek retired households recorded the highest availability (340 g/p/d), whereas Finnish households of the category "others" present the lowest availability (86 g/p/d).

Greece recorded the highest mean availability values of all three locality categories, Finland presents the lowest values. The lowest availability is presented by Finnish secondary educated households (110 g/p/d), whereas Greek elementary educated households present the highest availability (292 g/p/d).

Bread and Rolls

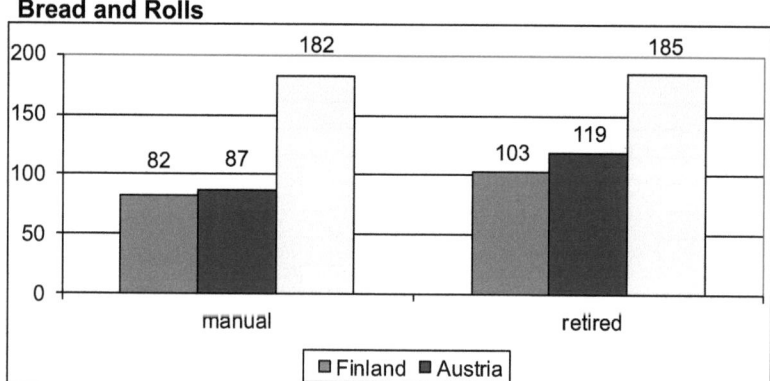

Figure 49. Daily Availability of bread and rolls, by occupation (g/p/d)

Greece presents the highest daily availability values of all household composition categories. Greek elderly single households recorded the highest availability (216 g/p/d), whereas Finnish elderly households with two members recorded the lowest availability (96 g/p/d).

Austrian higher educated households recorded the lowest availability of bread and rolls (80 g/p/d), whereas Greek elementary households recorded the highest availability (199 g/p/d).

Greece recorded a higher mean availability (167 g/p/d) of bread and rolls than Finland (88 g/p/d) and Austria (91 g/p/d).

Rural and semi-urban Finnish households present lower availability values than Austrian households. Austrian urban households recorded an availability of 88 g/p/d, Finnish urban households recorded an availability of 89 g/p/d.

Beverages

Non alcoholic beverages

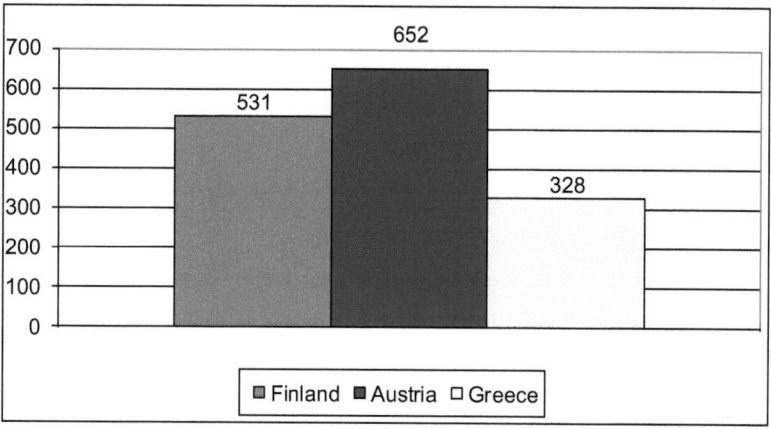

Figure 50. Daily availability of non alcoholic beverages – mean (ml/p/d)

Austria presents the highest availability values of all occupational categories. Greece presents the lowest availability values of the categories manual, non-manual, retired, and unemployed and Finland the lowest availability of the category "others". The lowest availability of non alcoholic beverages was recorded by Greek unemployed households (270 ml/p/d), whereas the highest availability was recorded by Austrian retired households (763 ml/p/d).

Elementary educated Austrian households present the highest daily availability of non alcoholic beverages (684 ml/p/d), secondary educated Greek households present the lowest daily availability (320 ml/p/d).

FOOD AVAILABILITY IN EUROPE

Austria recorded the highest mean availability of non alcoholic beverages (652 ml/p/d), Greece recorded the lowest availability (328 ml/p/d).

Austrian urban households present the highest availability value (675 ml/p/d) of all locality categories; Finnish rural households show the lowest daily availability (519 ml/p/d).

Alcoholic beverages

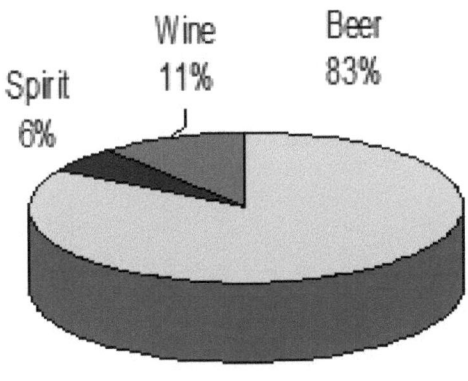

Finland

Figure 51a. Distribution of alcoholic beverages in North, Central, South Europe

FOOD AVAILABILITY IN EUROPE

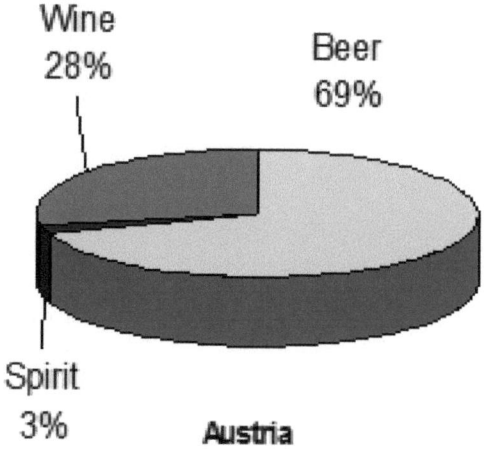

Figure 51b. Distribution of alcoholic beverages in North, Central, South Europe

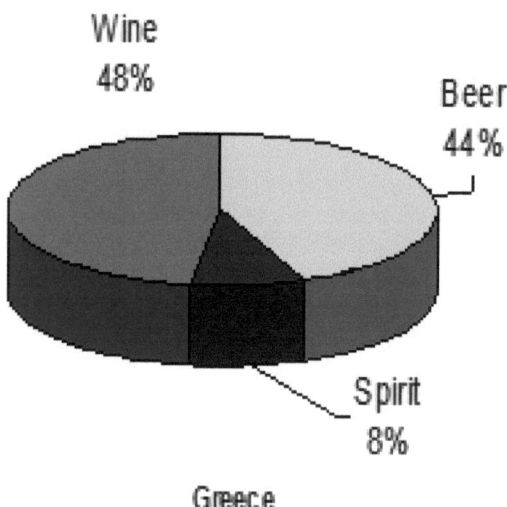

Figure 51c. Distribution of alcoholic beverages in North, Central, South

Austria presents the highest mean availability of alcoholic beverages (171 ml/p/d), Greece presents the lowest availability (48 ml/p/d).

Austrian households recorded higher daily availability values of all locality categories. The highest value was recorded by Austrian semi-urban households (200 ml/p/d), the lowest value was recorded by Finnish rural households (80 ml/p/d).

Austrian households show the highest mean availability of all occupation categories, Greece presents the lowest values of all occupation categories. The highest daily availability was recorded by Austrian retired households (230 ml/p/d), the lowest availability was recorded by Greek households of the category "others" (23 ml/p/d).

Greece presents the lowest daily availability values of all education categories. The lowest availability was recorded by Greek higher educated households (44 ml/p/d), whereas the highest availability was recorded by Austrian secondary educated households (174 ml/p/d).

Austrian households present the highest availability of all household composition categories.

The highest mean availability was recorded by Austrian adult single households (244 ml/p/d); the lowest availability of alcoholic beverages was recorded by Finnish elderly single households (35 ml/p/d).

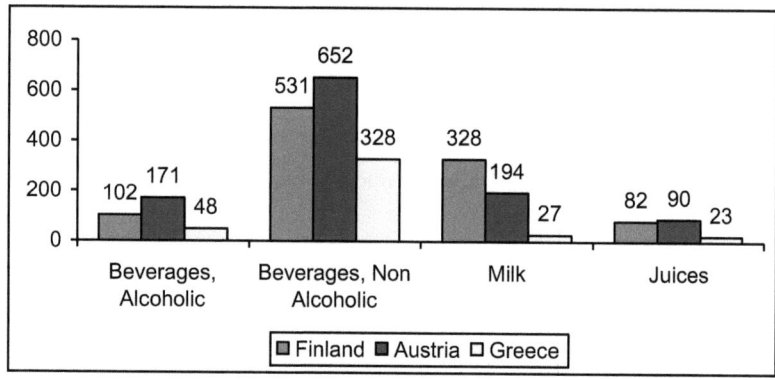

Figure 52. Mean daily availability in North, Central, South Europe (ml/p/d)

Milk Products

Finland presents the highest mean availability values of all household composition categories, whereas Greece presents the lowest values. The highest daily availability was recorded by Finnish elderly 2-member households (165 g/p/d), the lowest availability was recorded by Greek adult 2-member households (30 g/p/d).

Finnish households with retired heads recorded the highest daily availability (164 g/p/d), Greek "manual" households recorded the lowest availability (20 g/p/d).

Finland shows the highest availability values in all locality categories. In Finland rural households present the highest daily availability (136 g/p/d), whereas the Austrian rural households show the lowest mean availability (54 g/p/d).

Finland presents the highest daily availability values in comparison with Greece and Austria.

Finnish households with heads of elementary education present the highest availability (142 g/p/d), Greek "elementary educated" households show the lowest availability (22 g/p/d).

Finland presents the highest mean availability (131 g/p/d), Greece presents the lowest availability (27 g/p/d).

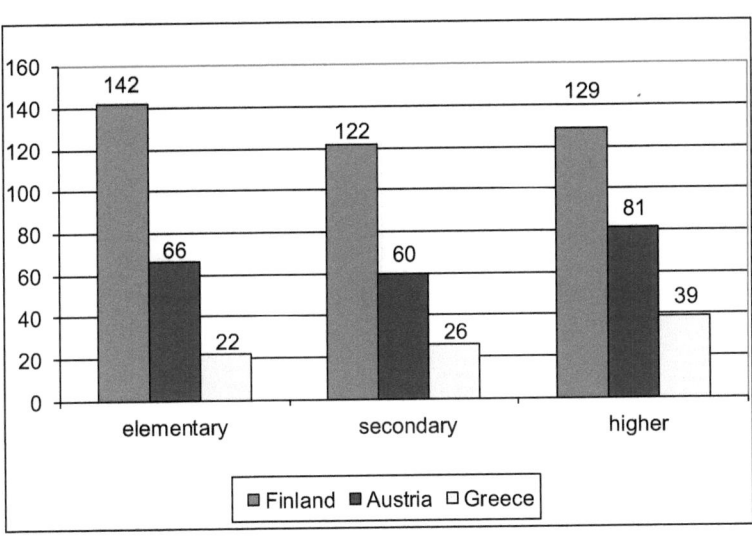

Figure 53. Mean daily availability of milk products, by education (g/p/d)

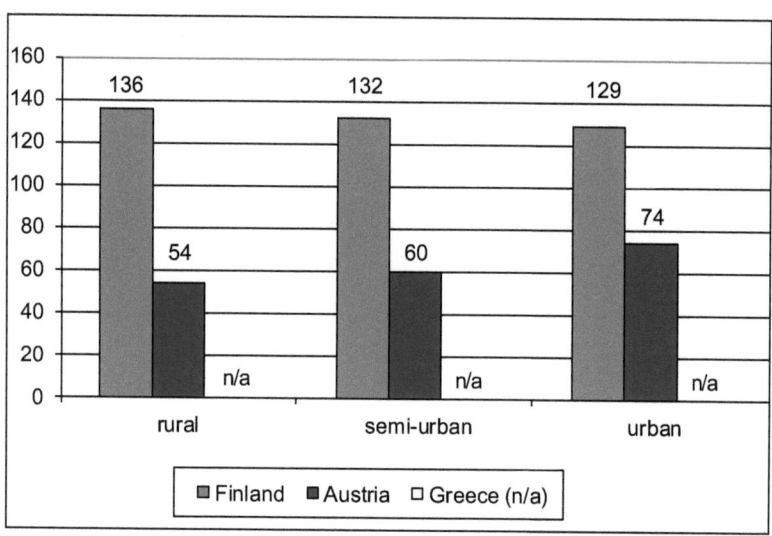

Figure 54. Mean daily availability of milk products, by locality (g/p/d)

Fish and Seafood

Greece presents the highest availability of fish and seafood (45 g/p/d), Austria presents the lowest availability (9,3 g/p/d).

Finnish rural households recorded the highest mean availability (34 g/p/d) whereas Austrian rural households recorded the lowest daily availability (7,88 g/p/d).

Austrian elementary educated households show the lowest mean availability (7,07 g/p/d); Greek elementary educated households present the highest mean availability (49 g/p/d).

FOOD AVAILABILITY IN EUROPE

The highest daily availability of fish and seafood (59 g/p/d) was recorded by Greek households of retired heads; the lowest availability was recorded by Austrian households of heads of the category "manual worker" (6,84 g/p/d).

Austrian adult single households recorded the lowest daily availability (8,72 g/p/d) in comparison to Greece (50 g/p/d) and Finland (38 g/p/d).

Total Added Lipids

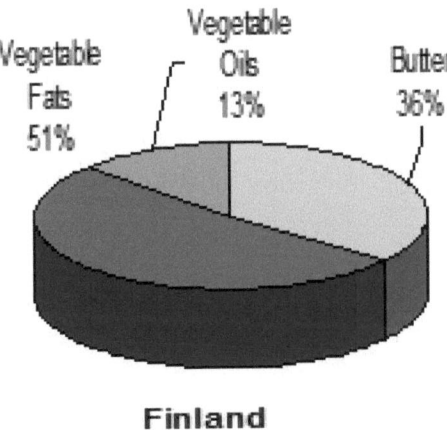

Figure 55a. Distribution of total lipids in North, Central, South Europe

FOOD AVAILABILITY IN EUROPE

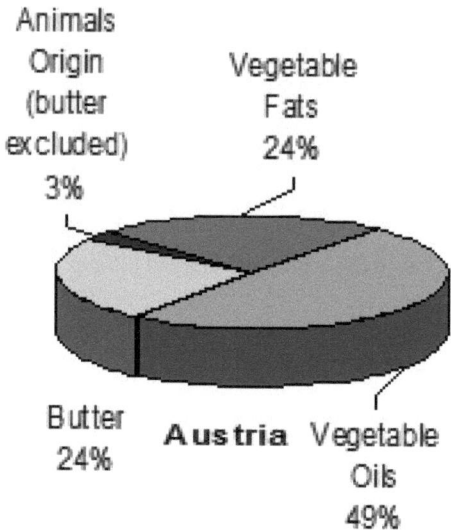

Figure 55b. Distribution of total lipids in North, Central, South Europe

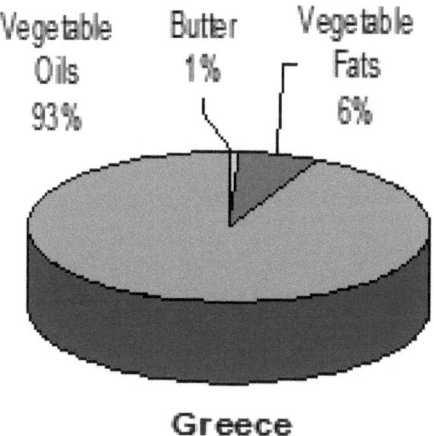

Figure 55c. Distribution of total lipids in North, Central, South Europe

Greece presents the highest mean availability of total added lipids (84 g/p/d), whereas Finland shows the lowest daily availability (31 g/p/d).

Greek elementary households recorded the highest availability (94 g/p/d), Finnish higher educated households recorded the lowest mean availability (25 g/p/d).

Austrian rural and semi-urban households recorded an availability of 44 g/p/d, whereas Finnish urban households recorded an availability of 28 g/p/d.

Finnish adult single households show the lowest daily availability (30 g/p/d), Greek elderly single households present the highest mean availability (125 g/p/d).

The lowest availability of total added lipids (19 g/p/d) was recorded by the occupation category "others" in Finland; the highest daily availability (96 g/p/d) was recorded by the category "others" (96 g/p/d) in Greece. Olive oil is important not solely for its own health benefits; it is also associated with the consumption of large quantities of vegetables in the form salads and equally large quantities of legumes in the form of cooked food (Trichopoulou A. and Lagiou P., 1997b, p.384). In Spain and Greece, the consumption levels of butter and animal fat are very low – these foods are not traditional in the Mediterranean countries (Prättälä R.S. et al., 2003, p.127). A lot of fat and oil is lost by roasting and frying or thrown away with salad marinades etc (Payer H. et al., 2000, p.7).

Sugar Products

Austrian households recorded a higher availability of sugar products (39 g/p/d) than Finland (25 g/p/d). Austrian urban households present the highest daily availability (47 g/p/d); the lowest daily availability was recorded Finnish rural households (23 g/p/d).

The highest mean availability is presented by households of the Austrian category "elementary" (48 g/p/d), the lowest daily availability is presented by Finnish category "elementary" (22 g/p/d). Austrian households with retired heads show the highest mean availability (47 g/p/d), whereas Finnish households of the categories "retired" and "unemployed" recorded the lowest availability (20 g/p/d).

Austrian elderly single households show the highest mean availability of sugar products (60 g/p/d). Finnish elderly households with two members recorded the lowest availability (19 g/p/d).

Sugar

Finnish adult single households recorded an availability of 23 g/p/d, whereas Austrian elderly single households show a mean availability of 74 g/p/d.

The highest availability was recorded by Austrian households of higher education (15 g/p/d), whereas the highest availability was recorded by Austrian households of elementary education (50 g/p/d).

Austrian households with retired heads present a daily availability of 59 g/p/d. Finnish households with heads of the category "others" recorded an availability of 16 g/p/d.

Austrian households present the highest availability of sugar (35 g/p/d), whereas Greek households recorded the lowest mean availability (26 g/p/d).

Finnish urban households recorded the lowest mean availability (23 g/p/d), whereas the Austrian rural households recorded the highest mean availability (46 g/p/d).

Juices

Austrian urban households present the highest daily availability of juices (112 ml/p/d). Finnish rural households show the lowest daily availability (70 ml/p/d).

Austrian households present the highest mean availability of juices (90 ml/p/d). Greek households present the lowest mean availability (23 ml/p/d).

Greece recorded the lowest availability values of all occupation categories, whereas Austria presents the highest availability values. The highest daily availability was recorded by Austrian unemployed households (124 ml/p/d). The lowest mean availability was recorded by Greek retired households (16 ml/p/d).

Greek households with heads of "elementary education" recorded the lowest (15 ml/p/d), whereas Austrian households with heads of "higher education" (117 ml/p/d).

Greek elderly 2-member households show the lowest daily availability (10 ml/p/d), whereas Austrian adult single households present the highest mean availability of all household composition categories (127 ml/p/d).

Bakery products

Finnish elderly single households recorded the highest daily availability (81 g/p/d), whereas Greek adult 2-member households recorded the lowest daily availability (22 g/p/d).

Finnish and Austrian "retired" households recorded an availability of 68 g/p/d; Greek "unemployed" households recorded an availability of 12 g/p/d.

Greek "elementary" and "secondary" households recorded a daily availability of 17 g/p/d, whereas Austrian "higher" educated households recorded a mean availability of 68 g/p/d.

The highest daily availability of all locality categories was recorded by Austrian urban households (71 g/p/d). Finnish rural and urban households recorded an availability 53 g/p/d.

Austrian households present the highest mean availability of bakery products (62 g/p/d), whereas Greek households present the lowest mean availability (18 g/p/d).

Cereals and Cereal Products

Austrian households present the highest mean availability (89 g/p/d), whereas Greek households present the lowest mean availability (18 g/p/d).

The highest daily availability of cereals and cereal products was recorded by Austrian urban households (109 g/p/d), whereas the lowest daily availability was recorded by Finnish urban households (22 g/p/d).

Austrian households with heads of higher education recorded an availability of 121 g/p/d, whereas Greek households with heads of secondary education recorded an availability of 16 g/p/d.

The lowest availability of all occupation categories was recorded by Greek unemployed households (16 g/p/d), whereas the highest availability of cereal and cereal products was recorded by Austrian unemployed households (123 g/p/d).

Austrian adult single households recorded an availability of 121 g/p/d, Finnish and Greek adult 2-member households recorded a daily availability of 22 g/p/d.

Flour

Austrian households present the highest mean availability (40 g/p/d), whereas Greek households present the lowest mean availability (22 g/p/d).

Elementary educated households show the highest daily availability, whereas higher educated households show the lowest daily availability in Austria, Finland and Greece.

The highest availability value was recorded by Austrian households of elementary education (98 g/p/d); the lowest daily availability value was recorded by Austrian and Greek households (15 g/p/d).

Austrian rural households show the highest availability (54 g/p/d), whereas Austrian and Finnish urban households present the lowest availability (23 g/p/d).

Greek non manual households show the lowest daily availability (16 g/p/d), Austrian retired households present the highest daily availability (60 g/p/d).

Greek adult single households recorded a lower daily availability (19 g/p/d) in comparison to Austrian and Finnish households, whereas Austrian elderly 2-member households recorded a higher mean availability (65 g/p/d).

Pasta

The lowest availability was recorded by Finnish elderly 2-member households (4,77 g/p/d), whereas the highest availability was recorded by Greek adult single and adult 2-member households (respectively 36 g/p/d).

Finnish "retired" households recorded the lowest daily availability (5,28 g/p/d), Greek "retired" and households of the category "others" recorded the highest availability (30 g/p/d).

The lowest availability value of all locality categories was recorded by Finnish elementary educated households (7,51 g/p/d), the highest availability was recorded by Greek elementary educated households (30 g/p/d).

Greece presents the highest availability of pasta (28 g/p/d), Finland presents the lowest availability (8,69 g/p/d).

Finnish rural households recorded the lowest daily availability (7,72 g/p/d), whereas Austrian urban households recorded (23 g/p/d).

Eggs

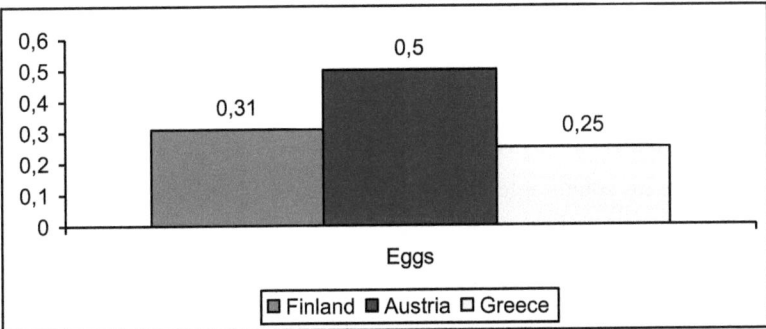

Figure 56. Mean daily availability of eggs (pieces/p/d)

Austrian households present the highest availability of eggs (0,5 pieces/p/d), Greek households present the lowest availability (0,25 pieces/p/d).

Austrian rural households recorded an availability of 0,56 pieces/p/d, whereas Finnish urban households recorded an availability of 0,3 pieces/p/d.

The lowest daily availability was recorded by Greek households with heads of "higher education" (0,22 pieces/p/d). The highest daily availability was recorded by Austrian households with heads of "elementary education" (0,59 pieces/p/d).

Greek "non-manual" households recorded an availability of 0,22 pieces/p/d, whereas Austrian "retired" households recorded an availability of 0,7 pieces/p/d.

Elderly single households recorded a daily availability of 0,8 pieces/p/d in Austria, whereas elderly 2-member households recorded a mean availability of 0,26 pieces/p/d in Greece.

Cheese

Austrian elderly 2-member households recorded an availability of 32 g/p/d, whereas adult single households recorded an availability of 73 g/p/d of cheese.

Greek "retired" and "others" households present the highest mean availability values of cheese (58 g/p/d), whereas Austrian households of the category "others" recorded the lowest mean availability (20 g/p/d).

"Higher educated" households recorded the highest daily availability values, whereas "elementary educated" households recorded the lowest daily availability in Austria, Finland, Greece.

Finnish households present the highest mean availability values of all locality categories. The highest availability was recorded by Finnish urban households (49 g/p/d), whereas the lowest availability was recorded by Austrian rural households (26 g/p/d). Greece presents the highest mean availability of cheese (53 g/p/d); Austria presents the lowest daily availability (28 g/p/d).

Milk

Finnish households present the highest availability of milk (328 ml/p/d); Austrian households present the lowest availability (194 ml/p/d).

Finnish "retired" households recorded the highest availability value (391 ml/p/d), Austrian "non manual" households present the lowest availability value (171 ml/p/d).

Austrian "higher educated" households show the lowest availability (155 ml/p/d), Finnish "elementary educated" households show the highest daily availability (378 ml/p/d).

Finnish households recorded the highest daily availability in rural areas (407 ml/p/d). Austrian urban households show the lowest daily availability (179 ml/p/d).

Finnish elderly single households show a mean availability of 417 ml/p/d; Greek adult 2-member households show a mean availability of 194 ml/p/d.

FOOD AVAILABILITY IN EUROPE

SURVEY YEAR	FINLAND 1998	AUSTRIA 1999 - 2000	GREECE 1998 - 1999
Eggs (pieces)	0.31	0,50	0.25
Potatoes (g)	111	100	137
Pulses (g)	1.3	N/A	14
Nuts (g)	1.6	N/A	4.7
Cereals (g)	206	303	253
Milk and milk products (g)	507	284	298
Cheese (g)	48	28	53
Meat and meat products (g)	148	182	149
Red meat (g)	52	65	100
Poultry (g)	12	23	39
Processed meat (g)	67	92	8.7
Vegetables (g)	123	142	271
Fresh vegetables (g)	103	114	248
Processed vegetables (g)	20	28	23
Fish and seafood (g)	30	9.3	45
Fruits (g)	157	192	306
Fresh fruit (g)	147	186	305
Processed fruit (g)	11	6.2	0.47
Fruit&vegetable juices (ml)	82	90	23
Lipids, added (g)	31	42	84
Animal fat (g)	11	12	0.80
Vegetable fat (g)	16	11	5.8
Vegetable oils (g)	3.5	22	77
Beverages alcoholic (ml)	102	171	48
Beverages non alcoholic (ml)	531	652	328
Soft drinks (ml)	81	116	65
Sugar and sugar products (g)	53	74	N/A

N/A: Non available
Source: The DAFNE databank

Table 23. Mean daily food availability in Finland, Austria, Greece

8.3.2. DAFNE members in comparison to mean DAFNE food availability

On the following pages the spider web **reference circle of radius 100%**, indicated in each figure, corresponds to **overall average (10 DAFNE members)** DAFNE daily availability. A data point **below 100%** indicates that the country recorded **lower** availability of that food group in comparison to the reference DAFNE average, and vice versa if the data point appears **above 100%**.

Values are calculated as arithmetic means of the mean availability values of ten DAFNE members: Austria, Belgium, Finland, Germany, Greece, Ireland, Italy, Norway, Portugal and the UK.

FOOD AVAILABILITY IN EUROPE

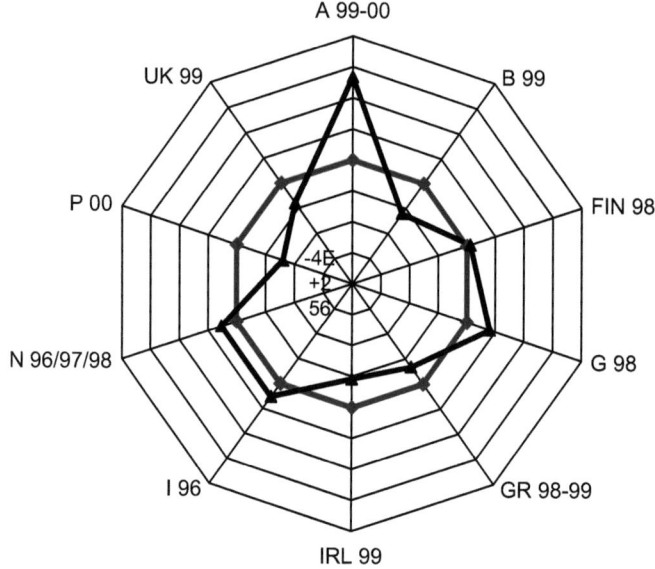

Figure 57. Eggs – mean

The reference circle corresponds to overall average availability of ten DAFNE members.

It must be noticed that Austria shows largest difference of daily availability of eggs to the DAFNE average.

FOOD AVAILABILITY IN EUROPE

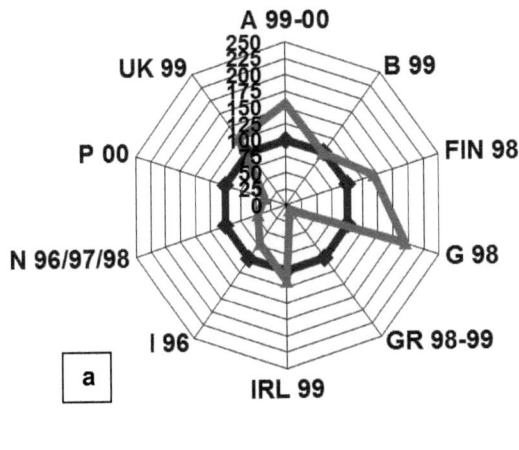

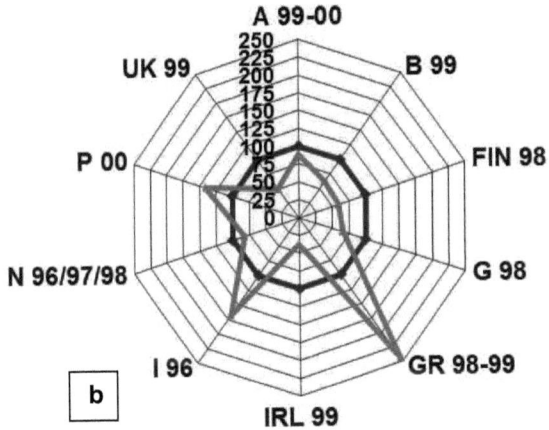

Figure 58. Lipids of (a) animal origin and (b) vegetable origin – mean

FOOD AVAILABILITY IN EUROPE

Regarding the availability of animal lipids, Austria, Germany, Finland, Ireland and the UK recorded higher availability values than the DAFNE average (reference circle). Regarding the availability of vegetable lipids, these countries show lower availability values than the DAFNE average (reference circle).

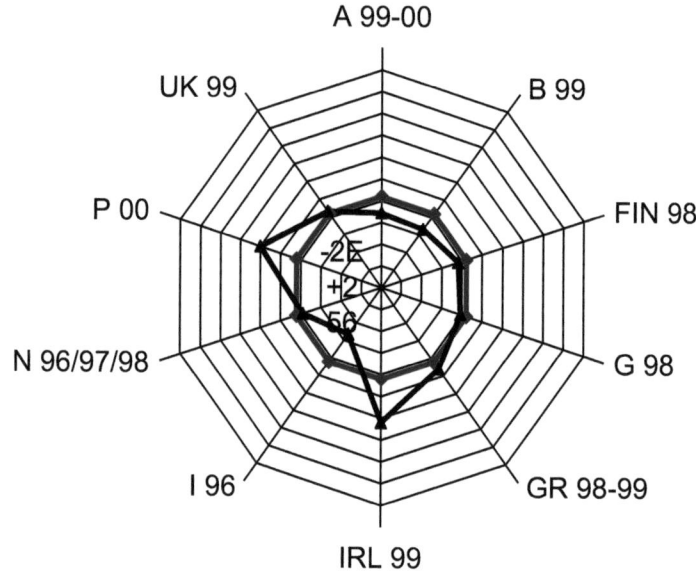

Figure 59. Potatoes – mean

The reference circle corresponds to overall average availability of ten DAFNE members.

Figure 59 indicates that Portugal and Poland recorded higher mean daily availability of potatoes than the DAFNE average.

FOOD AVAILABILITY IN EUROPE

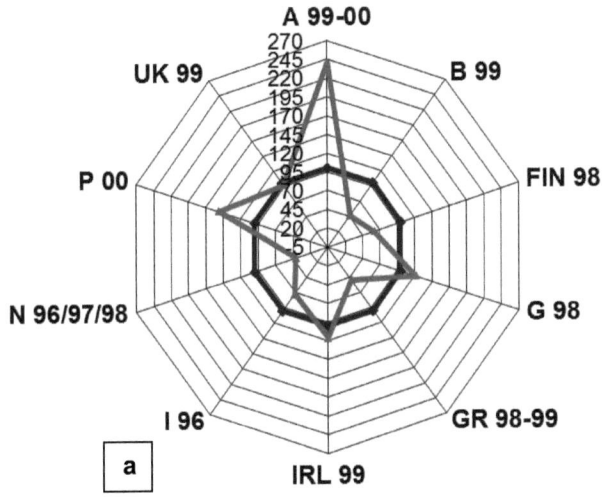

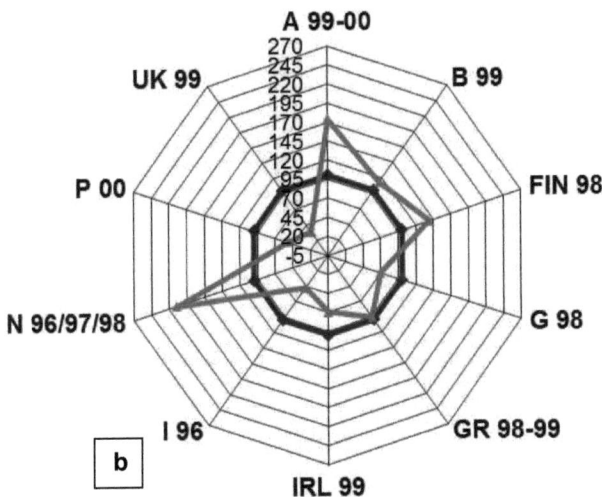

Figure 60. (a) Cereals and cereal products and (b) Flour – mean

FOOD AVAILABILITY IN EUROPE

The reference circle corresponds to overall average availability of ten DAFNE members.

Austria shows the largest difference of cereal and –products availability to the DAFNE average. Figure 60 b indicates that Norway, Austria, Belgium and Finland recorded higher availability values of flour than the DAFNE average.

Figure 61 a indicates that Austria show a lower cheese availability than the DAFNE average (reference circle). Figure 61 b indicates that Italy show the largest difference of pasta availability to the DAFNE average (reference circle).

FOOD AVAILABILITY IN EUROPE

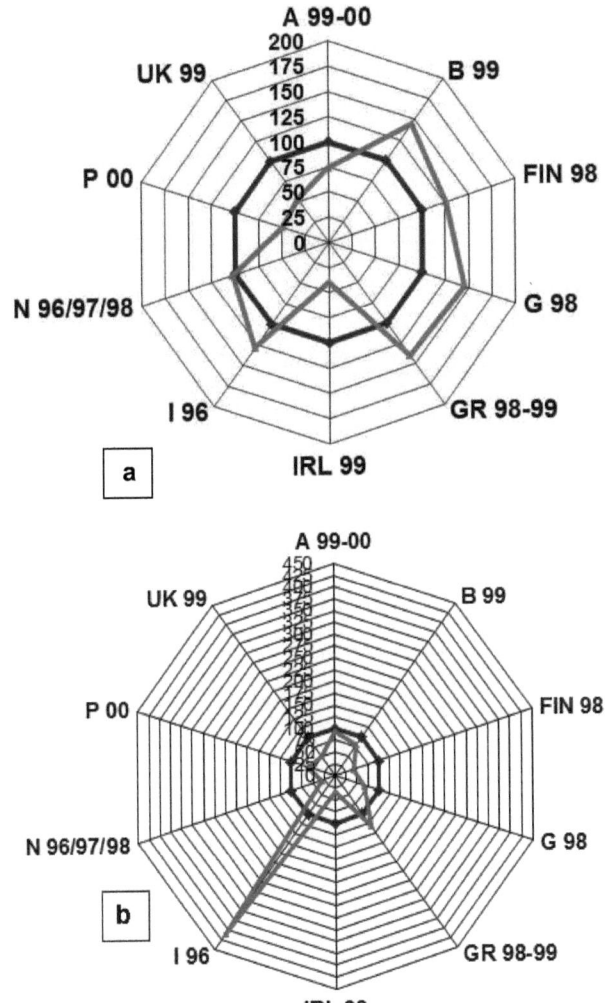

Figure 61. (a) Cheese and (b) Pasta – mean

FOOD AVAILABILITY IN EUROPE
8.3.3. DAFNE members in comparison to Austrian mean food availability

On the following pages the spider web **reference circle of radius 100%**, indicated in each figure, corresponds to the **Austrian** DAFNE average daily availability. A data point **below 100%** indicates that the DAFNE member recorded **lower** availability of that food group in comparison to Austria, and vice versa if the data point appears **above 100%**.

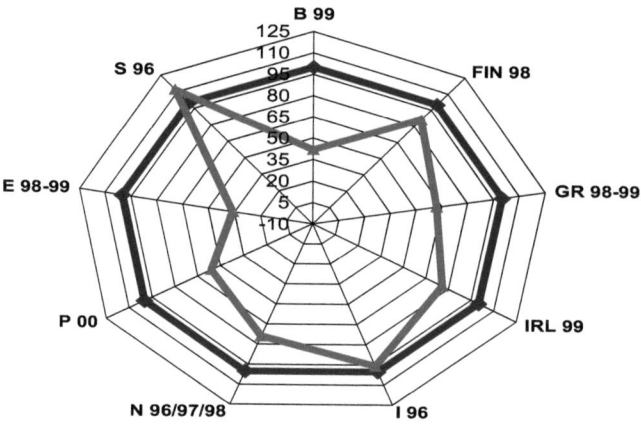

Figure 62a. Sugar by Elementary education

FOOD AVAILABILITY IN EUROPE

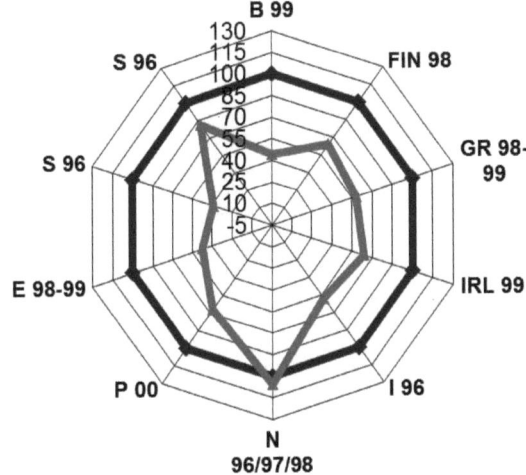

Figure 62b. Sugar by Secondary education

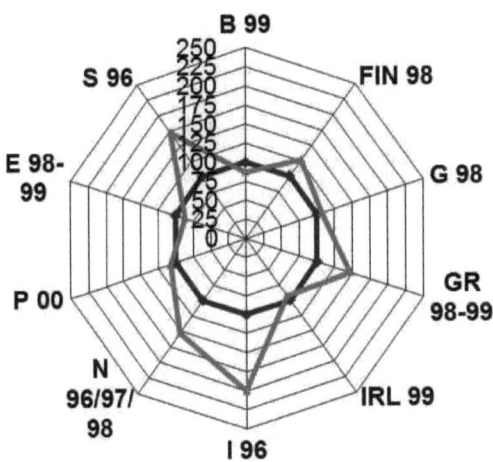

Figure 62c. Sugar by Higher education

The reference circle (figure (62 a-c.) corresponds to the Austrian daily availability. It has to be noticed that only Sweden recorded higher values of sugar availability than Austria in the category of elementary education and Italy of secondary education. With respect to households of higher education only two DAFNE members showed lower availability of sugar, namely Spain and Belgium.

The reference circle (figure 63 a-c.) corresponds to Austria. Austria shows the highest daily availability of cereals and cereal products of all DAFNE members in all three education categories.

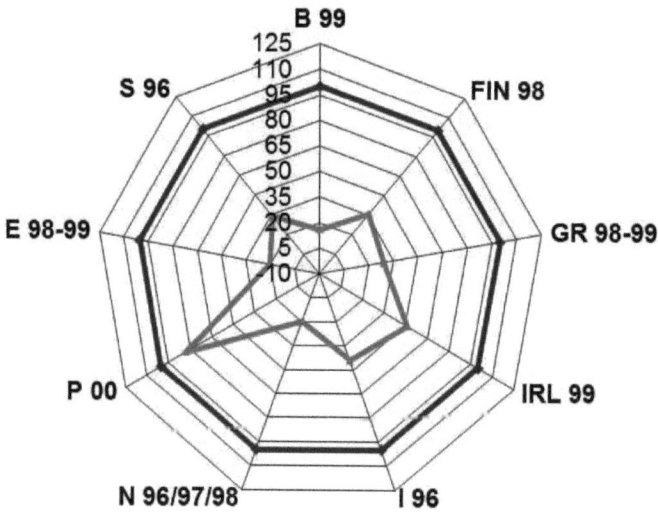

Figure 63a. Cereals and Products, by Elementary education

FOOD AVAILABILITY IN EUROPE

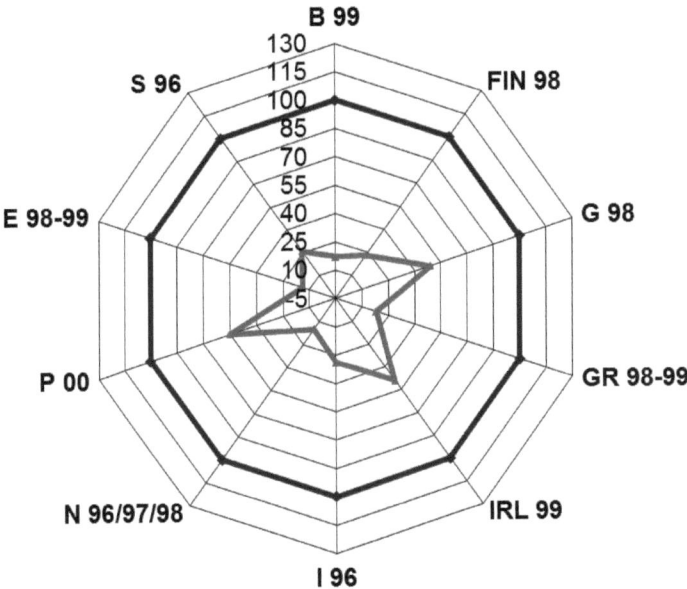

Figure 63b. Cereals and Products, by Secondary education

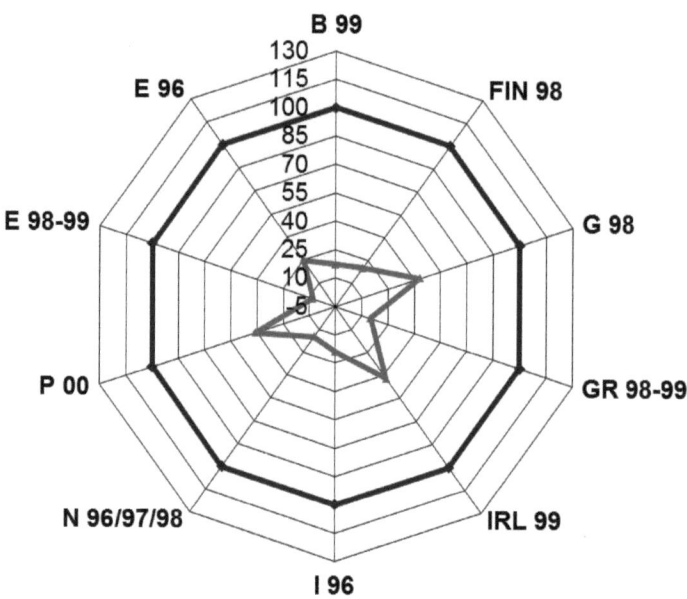

Figure 63c. Cereals and Products by Higher education

The reference circle corresponds to Austria. Austria shows the highest daily availability of cereals and cereal products of all DAFNE members in all three education categories.

FOOD AVAILABILITY IN EUROPE

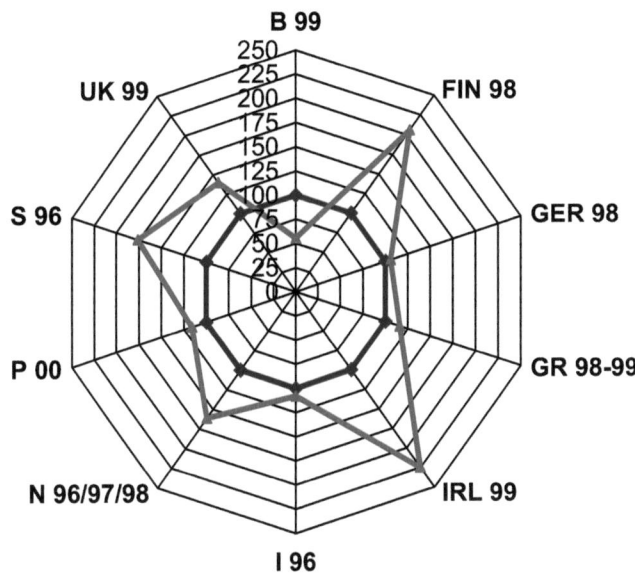

Figure 64a. Milk, by Manual household head

FOOD AVAILABILITY IN EUROPE

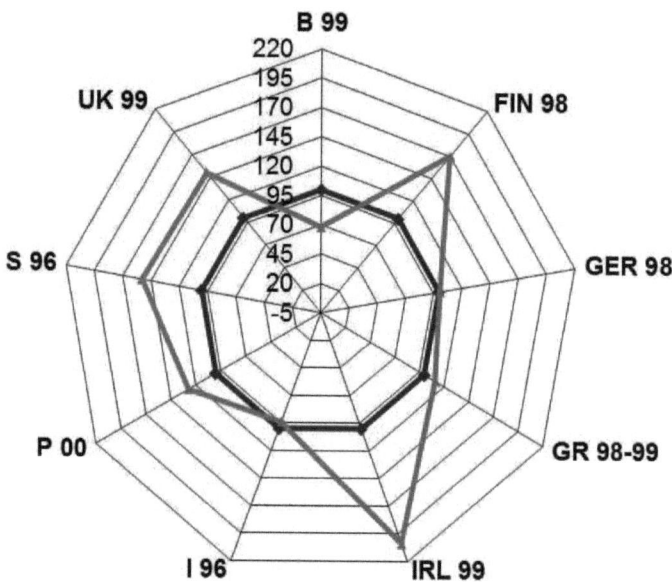

Figure 64b. Milk, by Unemployed household head

The reference circle corresponds to Austria. It must be noticed that only Belgium recorded lower availability values of milk than Austria in the categories manual and unemployed occupation.

FOOD AVAILABILITY IN EUROPE

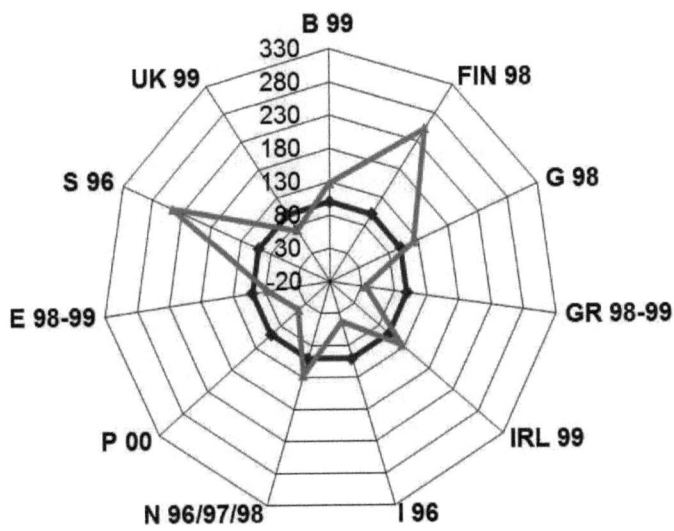

Figure 65a. Milk products in Rural areas

FOOD AVAILABILITY IN EUROPE

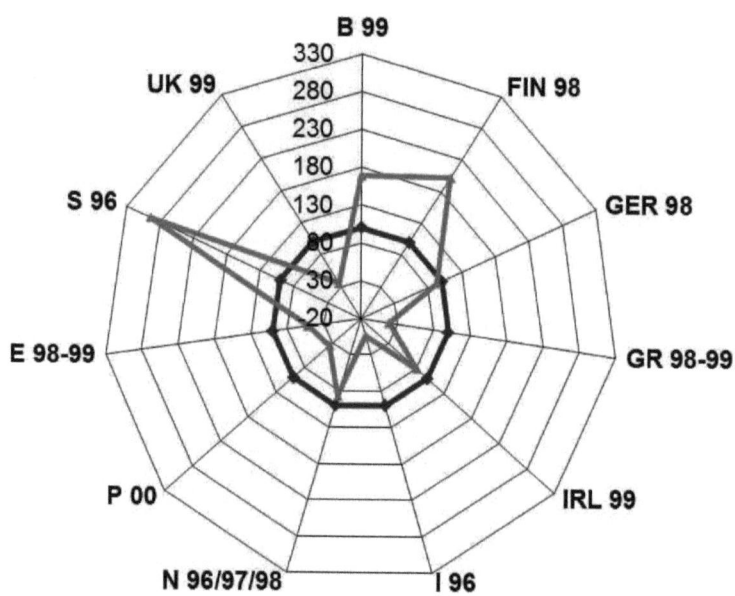

Figure 65b. Milk products in Urban areas

The reference circle corresponds to Austria. Figure 65 indicates that Sweden, Belgium, Finland and Germany have higher availability values of milk products in rural areas. The same pattern with the exception of Germany can be observed in urban areas.

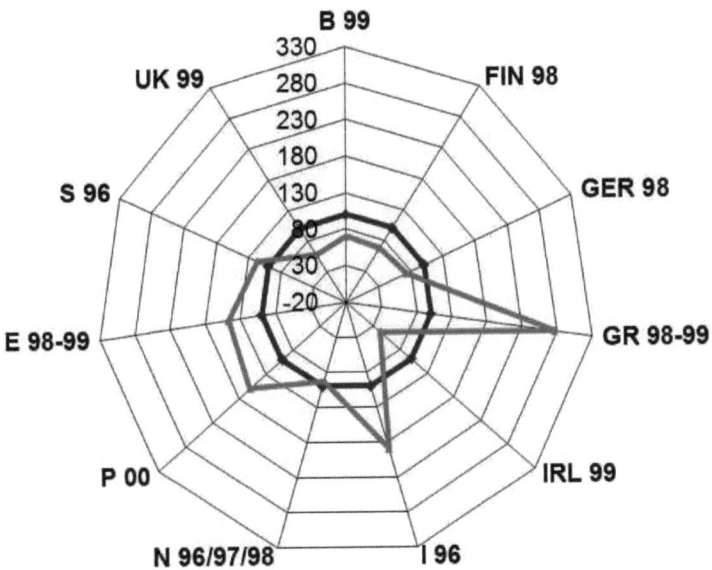

Figure 66. Lipids of vegetable origin – mean

The reference circle corresponds to Austria. The Mediterranean DAFNE members Spain, Portugal, Italy and Greece and the North European country Sweden show higher mean daily availability values vegetables lipids than Austria.

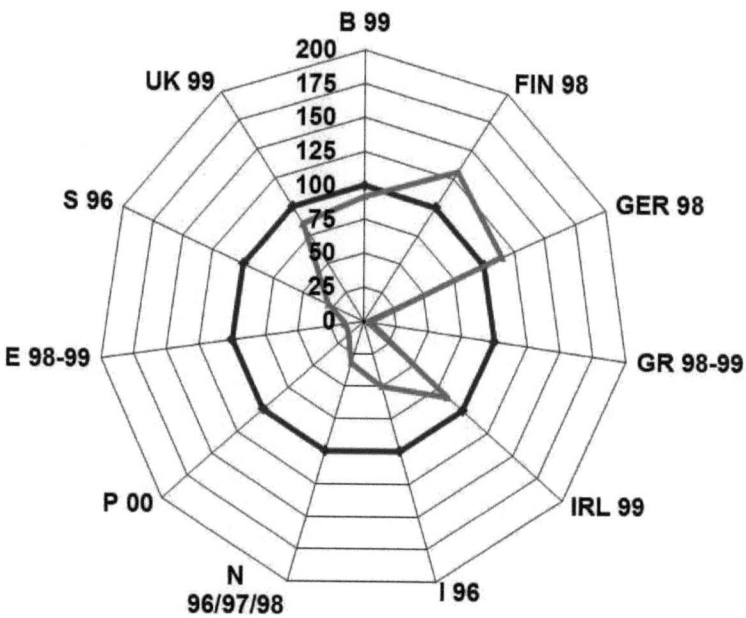

Figure 67a. Lipids of animal origin in Rural areas

FOOD AVAILABILITY IN EUROPE

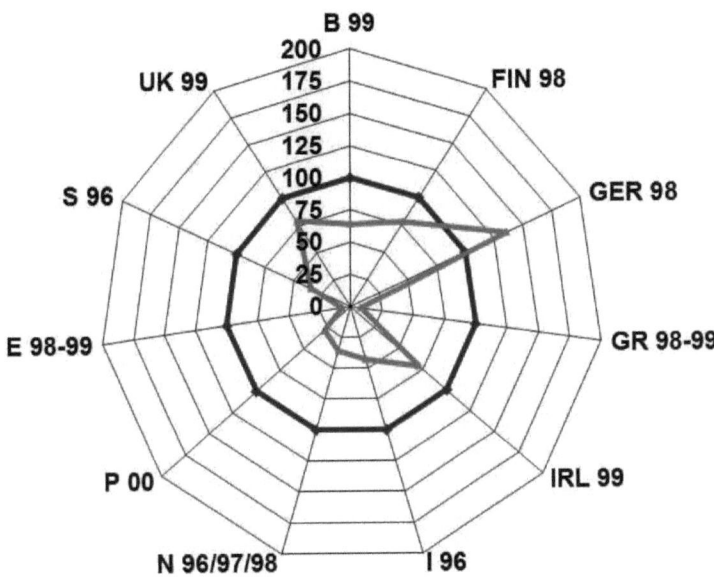

Figure 67b. Lipids of animal origin in Urban areas

The reference circle corresponds to Austria. In rural areas only Finland and Germany recorded higher availability values of animal lipids than Austria. In urban areas Germany shows a higher daily availability value than Austria.

FOOD AVAILABILITY IN EUROPE

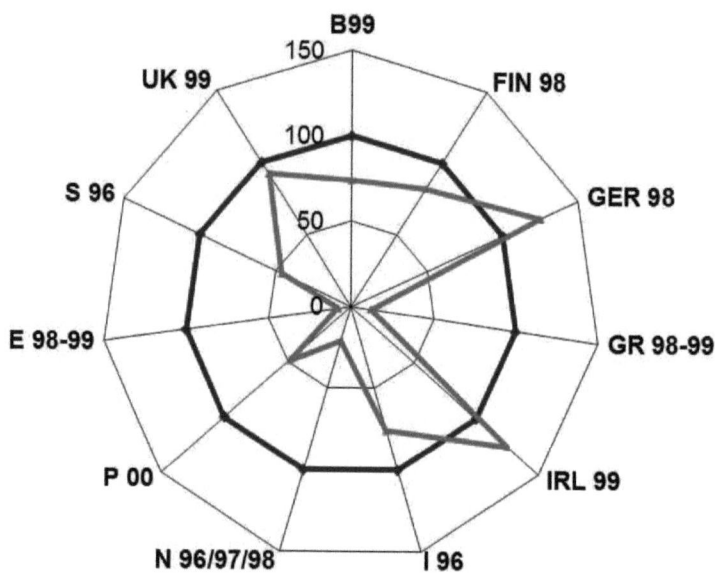

Figure 68a. Lipids of animal origin "adult-single" HH

FOOD AVAILABILITY IN EUROPE

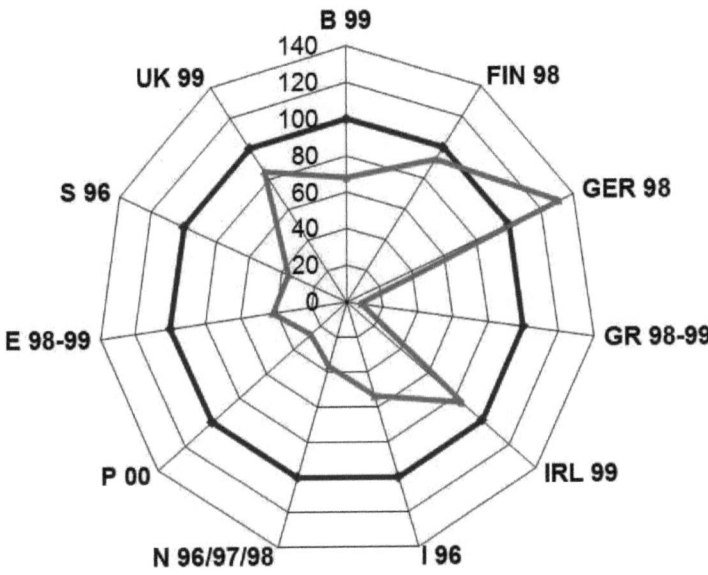

Figure 68b. Lipids of animal origin "adult-2-members" HH

The reference circle corresponds to Austria. It has to be noticed that only Germany shows higher availability values of animal lipids in the household composition categories adult-single and adult-two-member households.

FOOD AVAILABILITY IN EUROPE

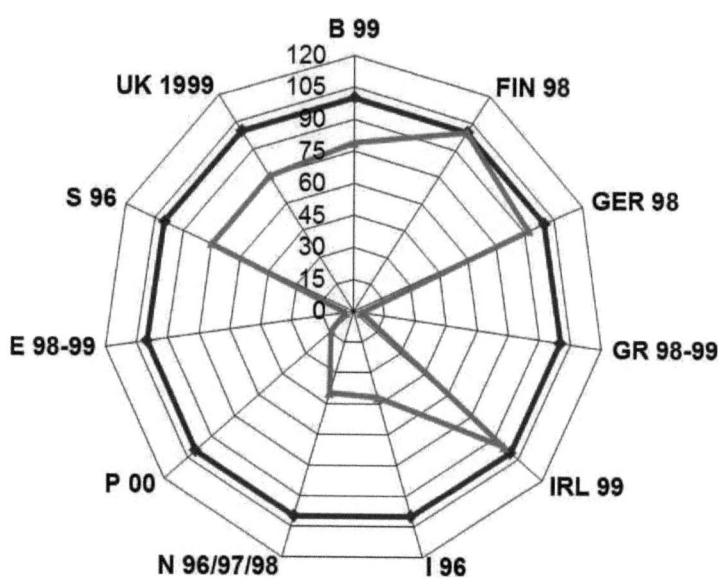

Figure 69a. Lipids of animal origin "elderly-single" HH

FOOD AVAILABILITY IN EUROPE

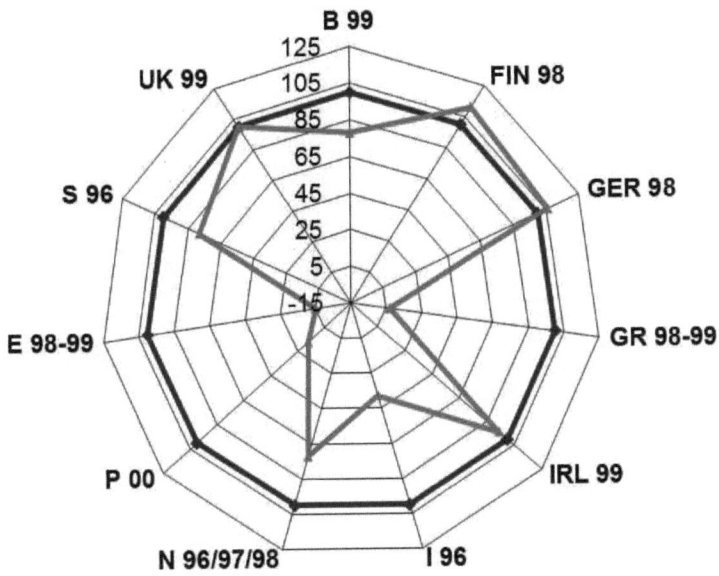

Figure 69b. Lipids of animal origin "elderly-2 members" HH

The reference circle corresponds to Austria. The Mediterranean countries Greece, Spain and Portugal show the largest difference to the Austrian animal lipids in the household composition categories elderly-single and elderly-two- member households.

FOOD AVAILABILITY IN EUROPE

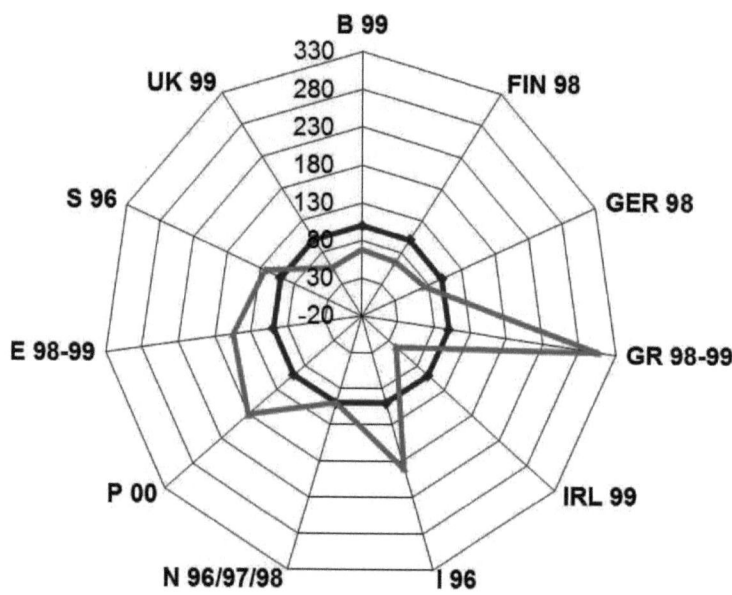

Figure 70a. Lipids of vegetable origin in Rural areas

FOOD AVAILABILITY IN EUROPE

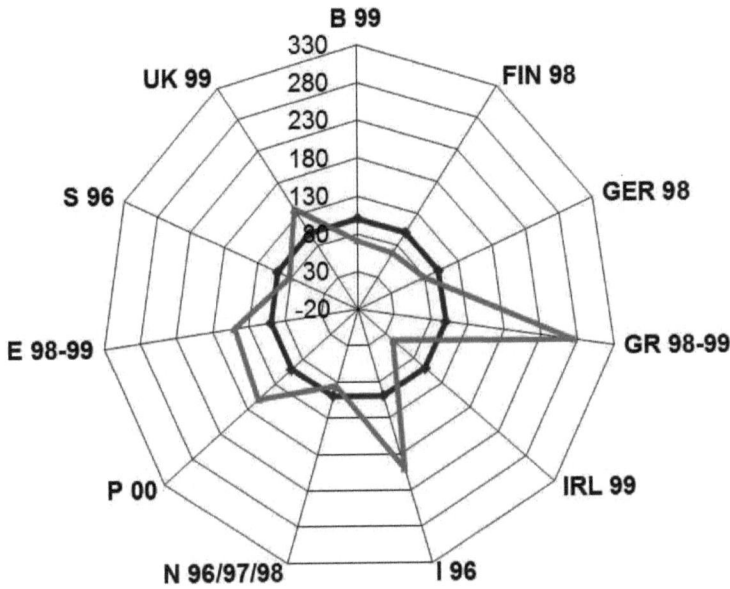

Figure 70b. Lipids of vegetable origin in Urban areas

The reference circle corresponds to Austria. All four Mediterranean countries Spain, Portugal, Italy and Greece show higher availability values of vegetable lipids than Austria in rural and urban areas.

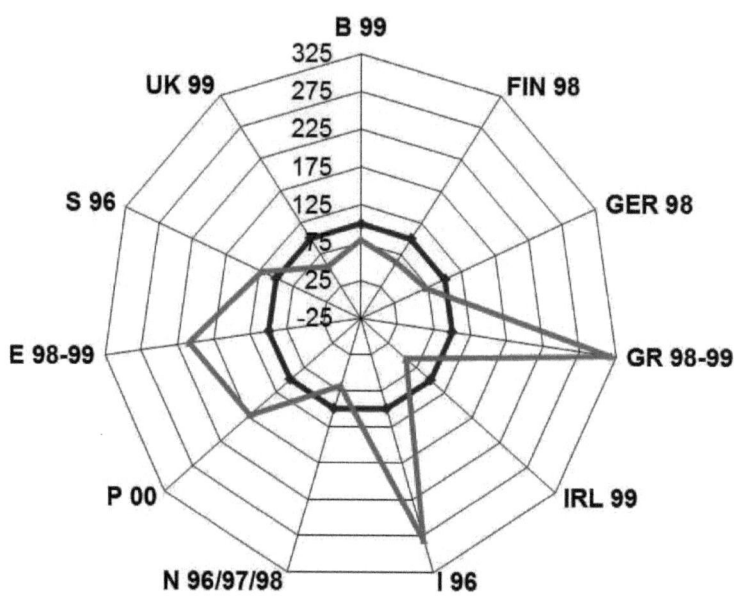

Figure 71a. Lipids of vegetable origin by "adult-single" HH

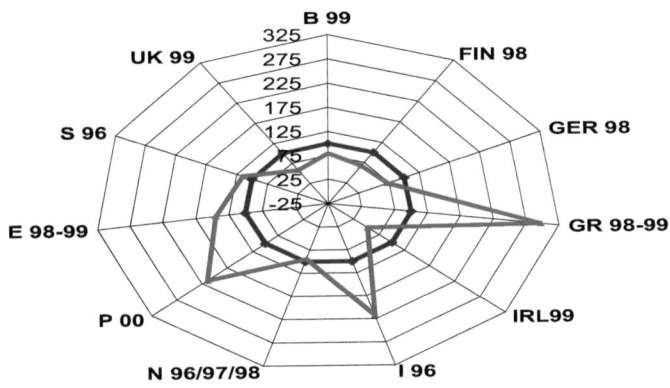

Figure 71b. Lipids of vegetable origin by "adult-2 members" HH

The reference circle corresponds to Austria. The North European countries (except Sweden) and the Central European countries recorded lower mean daily availability values of vegetable lipids than Austria in adult-single and adult-two-member households.

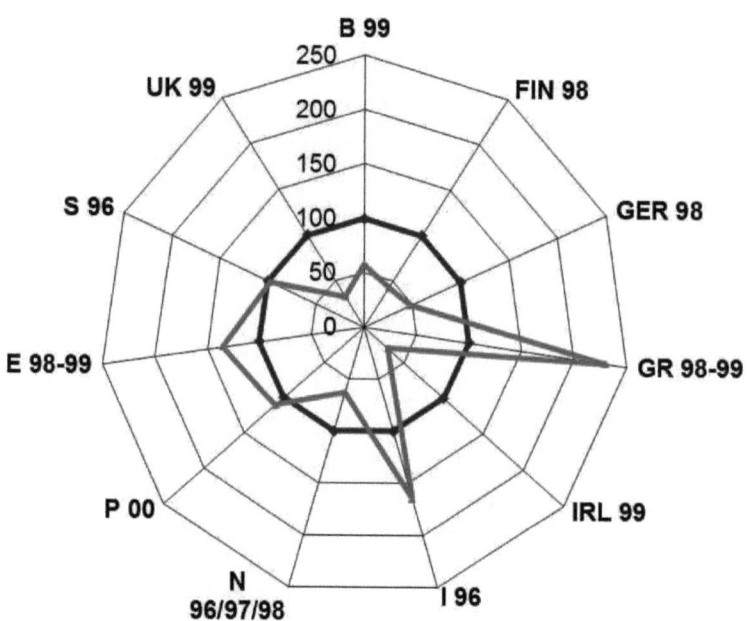

Figure 72a. Lipids of vegetable origin by "elderly-single" HH

FOOD AVAILABILITY IN EUROPE

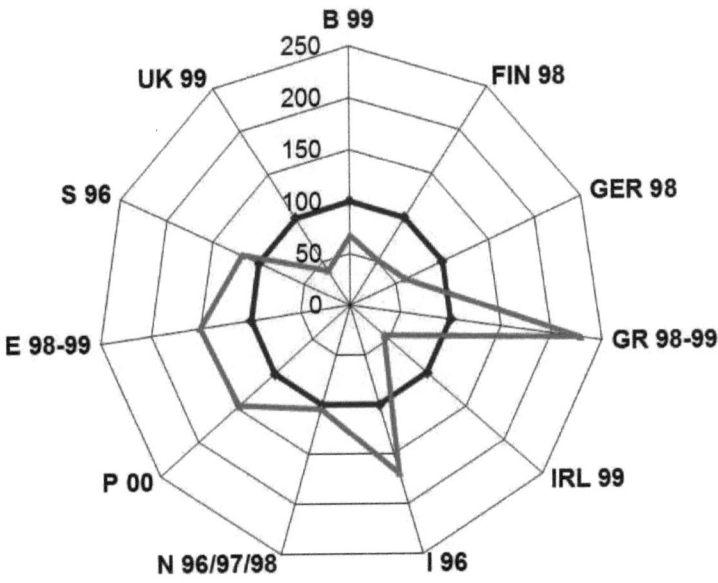

Figure 72b. Lipids of vegetable origin by "elderly-2 members" HH

The reference circle corresponds to Austria. The Mediterranean DAFNE countries show higher daily mean availability values of vegetable lipids than Austria in elderly-single and elderly-two-member households.

FOOD AVAILABILITY IN EUROPE

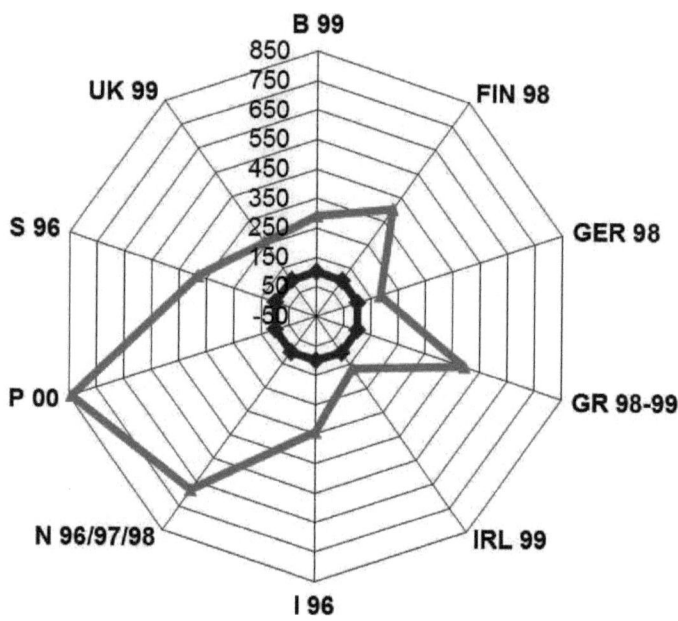

Figure 73a. Fish and Seafood by Retired household head

FOOD AVAILABILITY IN EUROPE

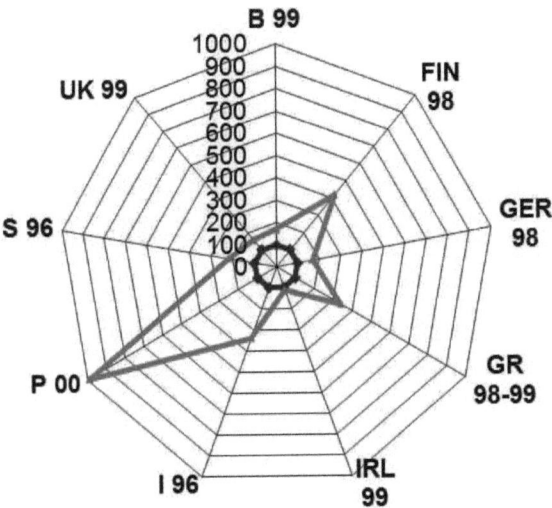

Figure 73b. Fish and Seafood by Unemployed household head

The reference circle corresponds to Austria.

Austria shows the lowest mean daily availability value of fish and seafood in comparison to the other DAFNE members in households with retired and unemployed household heads. Sweden recorded the highest daily availability in retired households, whereas in the category unemployed household heads the Portuguese households presented the highest value of fish and seafood availability.

9 DISCUSSION

9.1. HBS VERSUS INS

Household versus individual food recording

Family values are more appropriate than man-value for describing nutrient distribution within families. Man values are the ratio of intakes of groups of individuals and describe nutrient distribution in the sample as a whole. Family values on the other hand, are averages of the ratio of intakes of individuals and take into account the relative ranking of individuals. They describe more accurately than man-values the patterns of consumption within households where "big" eaters or "small" eaters tend to be clustered. Thus the HBS are appropriate, since they provide food availability information on a family/household level. However it is the family that influences directly the health of its members (Kanellou A., 1999, p.77).

9.2. HBS VERSUS FBS

Since HBS and FBS data are to a very large extent independent, a combination of information from both of these data sources can substantially reduce non – differential misclassification and improve the association with desirable, but not directly measured variable, which is the average individual dietary consumption (Lagiou P. and Trichopoulou A., 1999, p.331).

An important difference between FBS and HBS lies in the fact that the latter do not in general cover food consumed out-of-home. This may

DISCUSSION

explain a major part of the discrepancy between the data obtained from HBS and those derived from FBS. It can be anticipated that the discrepancy in the European developed countries where the opportunity for people to consume food away from the household is high. However, the data from the two sources may be close if the HBS take into account food consumed away from home in their measurements.

Certain kinds of food may not be covered by FBS because they are not included in national production statistics. Meats, such as those of game, wild animals as well as some crops may be excluded in this manner. Under conditions such as those prevailing in many developing countries, these meats may form a substantial part of the low consumption level of animal protein. Per capita food consumption data derived from HBS, multiplied by population numbers, could possibly provide the required production estimates (FAO, 1983, S.2-4).

9.3. HBS VERSUS FBS AND INS

The HBS takes up a position between the FBS and the INS. HBS and FBS allow between-country comparisons at a regular basis, but in moving from total population to household level, the HBS can provide a more detailed description of the dietary choices of the population, as well as of population subgroups (Naska A. et al., 2001b, p.1159-1160).

Trends of per-head-food-availability can be derivated from FBS, although these do not show the distribution in several population groups. Whereas results of HBS can be tabulated in subgroups of social, economical,

DISCUSSION

geographical and demographical factors, whichever, what kind of additional data are collected.

FBS show higher data than HBS, which are later collected in the "foodstuff chain". FBS represent useful information sources about trends and levels of food consumption (Becker W. and Helsing E., 1993, pp.60).

In a nutritional policy context, individual data on food consumption, energy and nutrient intakes provides the most appropriate information as it is segregated by age and sex, thus enabling the identification of high risk population groups.

In addition, individual dietary data may provide more suitable information that fits the needs of a national policy. However, household and national data may also be very useful in a nutritional policy context as they provide a useful estimate of food availability. Knowing food availability is a very important step when developing dietary guidelines, as average individual consumption levels will only be reached if they are available at a national and household level. If available levels of national food consumption are higher than the desirable individual levels, nutrition education may have an important role in the nutrition policy. In contrast, if apparent levels are lower than recommendations, promotion of supply or trade may have an important role in policy development. The level of food information required will depend on the scope of the intervention. If the aim is to promote dietary changes at the household level via education or price interventions, HBS may be the best dietary tool for monitoring changes. If the scope is to promote national food supply changes by legislation, agricultural incentives, or international trade, FBS would be the most appropriate tool.

DISCUSSION

While FBS and HBS can easily be compared between countries as, in principle, the reproducibility is high, individual data include several methods (food frequency questionnaires, 24 – hour recall, food records, etc.), which make comparisons more difficult. In general, when compared with home food consumption studies, HBS overestimates consumption, mainly because of storage, consumption of food by visitors, or under-recording of intakes.

The differences between the estimate of FBS, HBS, INS may be attributable to:
- (1) differences in methods; each methodology has its own advantages and disadvantages,
- (2) differences in the indicator measured (food availability, food consumption, food waste, etc), and
- (3) differences in the population group analysed.

FBS and HBS comprise the entire population and country, but INS only consider certain population groups and regions.

Usually food availability at national level estimates FBS are higher than HBS, and they are higher than the results from INS. If results from INS are greater than results from HBS, differences may be explained by the different age group categories (for instance, if the INS sample excludes children, who consume less fruits and vegetables, INS may show higher levels of fruit and vegetable consumption than HBS).

DISCUSSION

HBS results are higher than FBS for certain foods; the differences may be explained by the fact that certain products are mainly consumed in the home, but not institutions or restaurants.

The three levels of dietary data provide unique information about the availability and reported intake of food in populations and their comparisons are difficult to interpret because each represents different steps of the food chain. The understanding of their relation could play an important part when formulating, evaluating, and monitoring a nutrition policy in a country or region (Serra-Majem L. et al., 2003, p.78-79).

9.4. RECORDING PERIOD

The number of recording days plays an important role for validity and accuracy. A longer recording period is in general better, but has also its limitations. The participants get tired or realize what they eat, which may cause changes in the selection of food.

The day of the week chosen for the book keeping could also influence the results since the weekends very often differ from the weekdays in terms of food patterns.

A recording period should at least include one weekday and one weekend day. On the other hand, the individual's true intake can be considered as the mean for a large number of days. The HBS record purchases for at least seven days and no longer than one month (two months in some cases in Hungary) (Kanellou A., 1999, p.79).

DISCUSSION
9.5. THE INFLUENCE OF SOCIO-ECONOMIC PARAMETERS

Household income as a socioeconomic indicator reasons:
Firstly, income is a well established and important determinant of dietary quality, and affects directly a family's ability to afford and procure food. Secondly, household income was likely to capture the socioeconomic characteristics of all people living in the household (reflecting individual level incomes, and to some extent education and occupation) and therefore presumably embodied most of the within household socioeconomic processes influencing food choice (Turrell G. et al., 2004, p.210).

People with a higher socioeconomic level tend to show a higher consumption of vegetables, fruits and fibre products and a lower consumption of meat, meat products, and fats than people from a lower socioeconomic level (Sanchez-Villegas A. et al., 2003, p.927). Low-income households expend less than higher income households and it is more difficult to induce low-income households to increase their expenditures on fruits and vegetables (Blisard N. et al., 2004, p.23).

There is the assumption that that a lower intake of meat, cheese, milk and fat is associated with a higher socioeconomic level in Europe.

Possibly, class differences may explain differences in food choice. Subjects in a higher social class or better educated people have different attitudes toward health and healthy foods perhaps because highly educated people have a better knowledge about a healthy diet than subjects in a lower social class or less educated subjects. These differences can perhaps

DISCUSSION

be explained by the fact that higher social groups are more interested, self-conscious, and responsible about their health (Sanchez-Villegas A. et al., 2003, p.927).

The occupational level is a measure of social prestige. Indeed, the educational level determines the occupation and jointly with occupation they determine the income level.

On the one hand, occupation and education measure aspects of the same socio-economic position. On the other hand, there are plausible pathways through which education and occupation could have independent role in predicting health-related behaviours. The amount of education and knowledge individuals acquire can influence their lifestyle, the importance given to preventive health measures. Occupation is related to differential exposure to environmental risk factors and to psychological stress. It determines income and therefore, access to certain food products. At the same time, it generates social networks that can greatly influence behavioural health habits. Thus, both indicators measure different pathways through which socioeconomic position can have an independent effect on diet. It is reasonable to conceive that, for example, poor dietary habits acquired in youth can be added to poor dietary choices in restaurants of an industrial complex where healthy diet may not be promoted. On the contrary, someone with high education may have broader knowledge about diet and health and will probably choose healthier meals at restaurants with colleagues who might also be more predisposed towards healthier habits (Galobardes B. et al., 2001, p.338).

DISCUSSION

Parameters of eating patterns, food preferences and the acquisition of likes and dislikes are the results of learning mechanism in a specific socio-cultural context.

People with a high level of concern for health are likely to make several consistent lifestyle choices, which include regular exercise and reasonable food selection (Bellisle F., 1999, p.359).

Mediterranean diet

In the social connection, the function of health nutrition of the household systems is the most essential task for acting within a household for all people, at any time, any age and in any situation of life. The frequency of critical parameters is less important than the knowledge of the socio-economic characteristics of different population groups with higher than the average frequencies of high parameters, unconcerned of being already pathological (Kanellou A., 1999, p.78).

The Mediterranean diet could be described as the dietary pattern found in the olive oil growing areas. The late 1950s and early 1960s – before the fast food culture began influencing the nutritional habits – the occurrence of coronary heart diseases and cancer in Mediterranean countries was much lower than in other industrialised countries (Leonhäuser I.-U. and Dorandt St., 2004, p.I/31).

The traditional Mediterranean diet is characterized, among others, by relatively low importance of milk and dairy products. Liquid milk has never played such an important role in the Greek diet like in the Polish one. Due to the climatic conditions it was frequently preserved and consumed as yoghurt and cheese. Thus intake of milk was rather low, while the

DISCUSSION

consumption of cheese, mainly soft cheese (feta) and yogurt high (Sekula W. et al., 1997, p.3-15).

The life expectancy of the Mediterranean population was higher, although medical care did not meet western standards. The relatively good health of Mediterranean people is not only based on the diet but also on their culture, history and lifestyle. Prof. Trichopoulou remarked that the relaxing psychosocial environment, mild climatic conditions, preservation of the extended – family structure, and even the afternoon siesta habit in the Mediterranean region may play contributory roles.

Due to globalisation and social change the original Mediterranean lifestyle explored in the early 1960s does no exist any more in the Mediterranean countries (Leonhäuser I.-U. and Dorandt St., 2004, p.I/31). Meat was once expensive and rarely consumed, but fish consumption was a function of proximity to the sea (Trichopoulou A. and Lagiou P., 1997b, p.383-384).

	Total Added Lipids (g/capita/day)	Olive Oil (ml/capita/day)
Austria (1999),	42	3,24
Finland (1998)	31	0,28
Greece (1998)	84	74
Spain (1998)	45	33

Figure 74. Mean Daily Per Capita Availability of Total Added Lipids and Olive Oil

DISCUSSION

Dietary structures in EU countries are becoming increasingly similar, although differences have been found of the dietary structure of the Mediterranean countries compared to other European countries.

The fact that Greek cuisine includes a lot of olive oil, and Northern European countries do not, is not unconnected with the geographical distribution of olive trees. Many international differences in dietary pattern are the consequence of availability (and price) in the locality. Consumers and cultures progressively acquire a taste for products that, even following the geographical spread of availability in the modern world, international differences in tastes and preferences remain. In Northern Europe olive oil is widely available now, at prices which no longer exceed by very much those in Mediterranean countries, and consumption levels differ more because of tastes and preferences than they do because of price and availability now (Gil Jose M., 1995, p. 393).

Figures 75 and 76 show the life expectancy of women and men in the representative North, Central and South European countries namely Finland, Austria and Greece.
Greek men have a higher life expectancy at birth and at age of 60 than Finnish and Austrian men, however Greek women have a lower life expectancy.

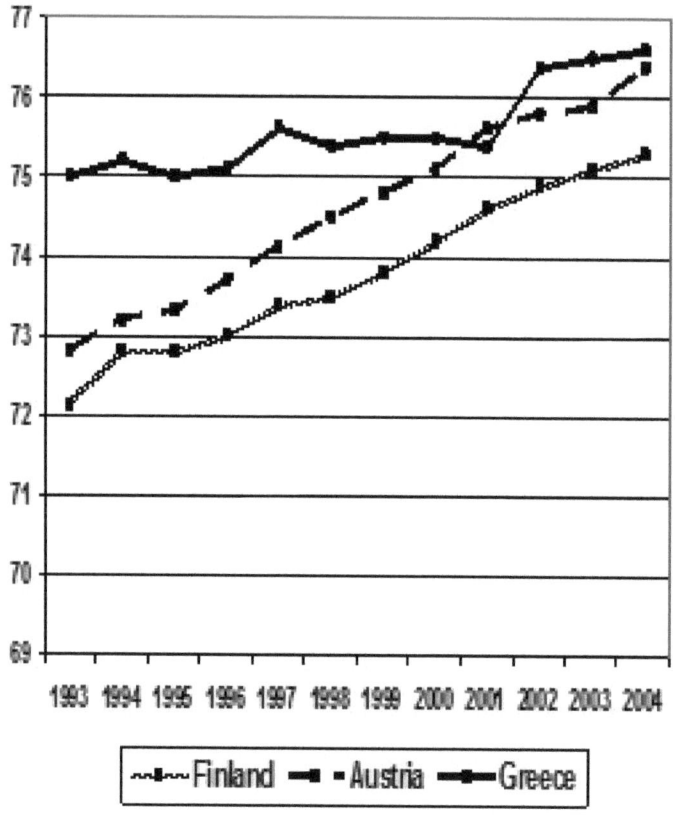

Figure 75a. Life expectancy at birth of men in North-, Central, South-Europe (age/year) (Eurostat 2007)

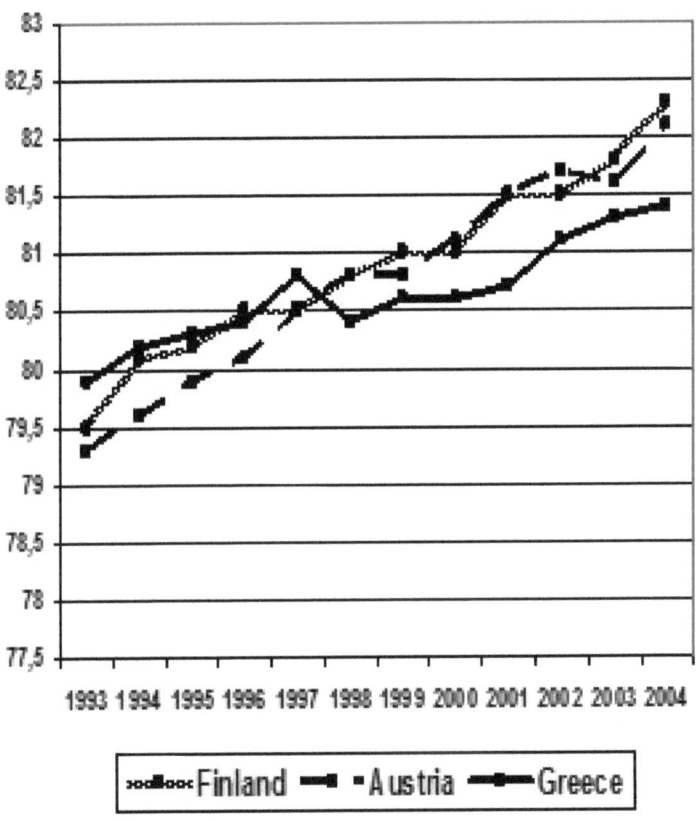

Figure 75b. Life expectancy at birth of women in North-, Central, South-Europe (age/year) (Eurostat 2007)

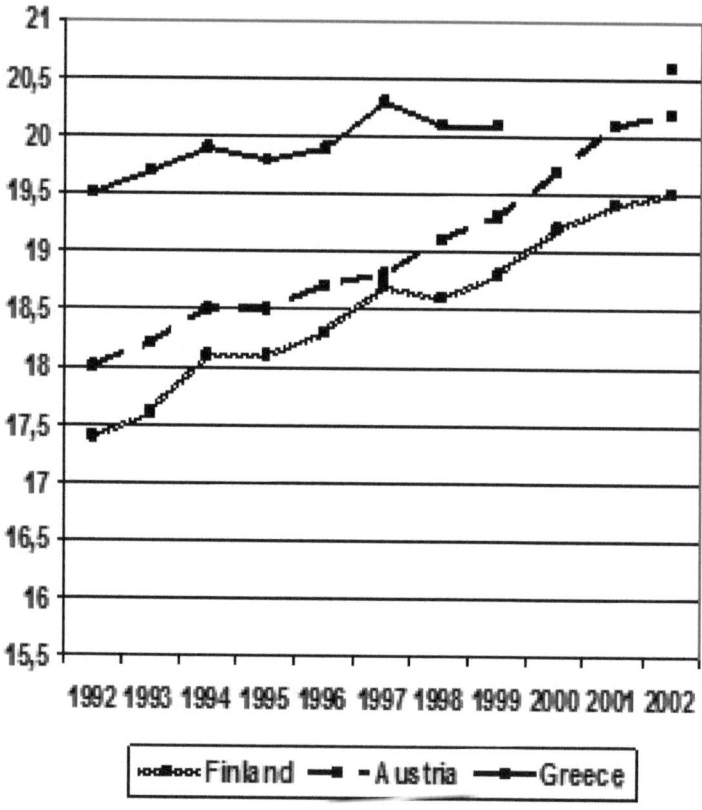

Figure 76a. Life expectancy at age 60 of men in North-, Central, South-Europe (age/year) (Eurostat 2007)

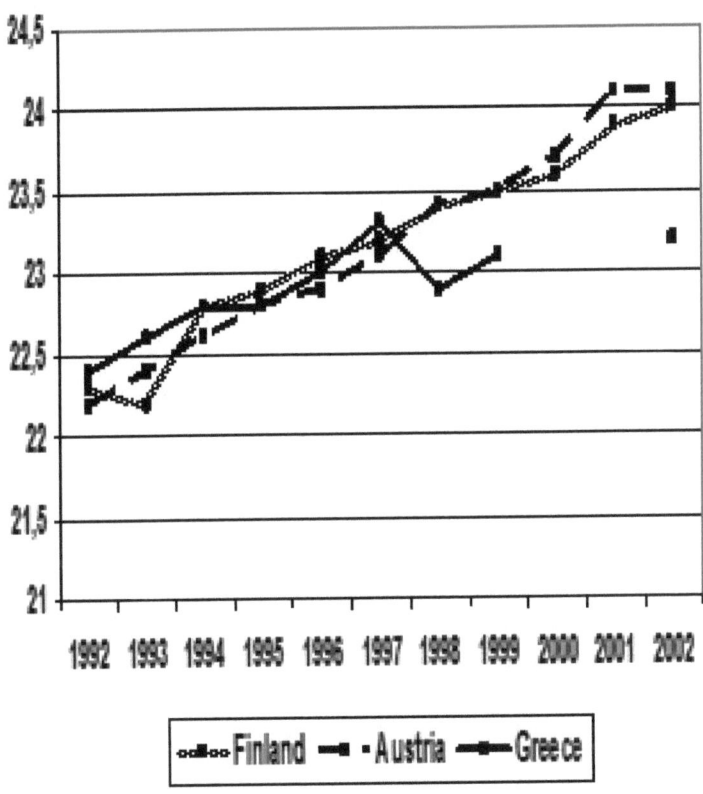

Figure 76b. Life expectancy at age 60 of women in North-, Central, South-Europe (age/year) (Eurostat 2007)

DISCUSSION
9.6. LIMITATIONS

9.6.1. Underreporting

Underreporting is another factor that has to be taken into account. Those belonging to the lower social classes may under-report their intake due to their lower education, whereas those belonging to higher classes may do it in order to respond to what social images expect from them (Sanchez-Villegas A. et al., 2003, p.927). Underreporting goes hand in hand with the risk of misclassification. Special attention for so-called "forgotten" foods (i.e., mostly soft drinks) (De Jong N., 1999, p.4)

Individuals may not wish to admit via surveys, even to themselves, exactly what they put in their mouths. The more scientists and media tell people that they should reduce the amount of fat in their diets, the more they tell how little of it they eat (Blundell J.E., 2000, p.3).

A further limitation of the HBS data for international comparison relates to different recording periods used in some of the countries (Naska A. et al., 2005, p.9), HBS provides a powerful, economical tool for obtaining information about the food availability of a wide cross-section of the population, in each country (Kanellou A., 1999, p.80).

It is important to keep in mind that a limitation of household availability data is that foods purchased and consumed away from home are not included. There is no doubt that these foods can contribute substantially to food and nutrient availability (Byrd-Bredbenner C. et al., 2000, p.198).

Furthermore, information on food losses and waste, food given to pets, meals offered to guests, use of vitamin and mineral supplements and the

DISCUSSION

presence of pregnant or lactating women in the household are not consistently collected.

The HBS data are collected at household level and therefore equal distribution of food and beverages within the household was assumed. Estimation of the individuals' intake requires the application of non-parametric modelling (Trichopoulou A. and Naska A., 2003a, p.28).

Another limitation of the HBS is that the results are obtained on the level of households and give evidence of the food consumption for the average person and it is impossible to determine the variability in food consumption among individuals. On the other hand, it must be emphasised that, compared with methods that investigate food consumption on the individual level, HBS is much less demanding financially and in terms of organisation, and provides at the same time reliable information that can be effectively applied in the area of nutritional epidemiology (Dofkova M. et al., 2001, p.1003).

Even data of good quality can be a source of error if they are derived from foods that are not clearly defined. The simple food name can be inadequate or ambiguous to those who are not closely acquainted with the local language and culture (e.g., "sweetbread").A common name may be misleading when the same name is used for different foods in different regions or when it is used for foods having different scientific names (e.g., "catfish"). Likewise, one may not recognize some terms used by people in other parts of the world or maybe even within the same country. The situation is further confused by homonyms, synonyms, identical brand names for different products, and culinary or technological terms.

DISCUSSION

As most databases employ different methods of identifying food, it is difficult to exchange data among countries, among organizations within the same country, or even among workers in the same institution (Ireland J.D., 2000, p.529).

9.6.2. The need for a unified HBS food coding system

Currently, only food availability data from HBS are comparable at the European level (DAFNE) (The EFCOSUM Group, 2002, p.S1). Based on different food classification systems, a simplified classification "EURO FOOD GROUP (EFG)" was proposed. EFG contains a list of 33 food groups, based on the study of different food classification systems (among others Eurocode 2, WHO GEMS regional diets, FAO food balance sheet, DAFNE) (EFCOSUM group, 2001, p.29). DAFNE with 45 groups makes more specific distinctions and Eurocode 2 with 13 main groups approximately 160 subgroups (Unwin Ian, 2000, p.1).

The aggregation and classification of food groups and items on a comparable level allows the comparison of the HBS data on food availability. A harmonization of terminology and the development of a compatible and comparable food coding system are of enormous importance. A unified questionnaire with a unified item classification would allow the comparison of food consumption in an easy, uniform and invariable way.

Based on the aggregation of HBS food codes some improvement, additions or changes are of importance.

DISCUSSION

- The distinction between white and whole-meal bread and cereal products would allow a better estimate of dietary fibre, vitamin and mineral intake.
- The creation of at least two cheese categories (e.g. fresh and hard) based on the variety in the fat content.
- To separate among the preserved fruits or vegetables, the frozen from the canned. There is a growing trend of buying frozen goods, which are not comparable with canned ones, in respect of nutrients, preservatives and, in some cases, sugar content.
- The out of-home consumption to be included with separate codes based on their main constituents. Meals at restaurants, sandwiches, salad, confectionery or chocolates, beverages and savoury snacks are worth to be recorded separately.
- The miscellaneous group not to include such a variety of different constituents in food.
- All the food purchased to be recorded in quantities and on their monetary cost (Kanellou A., 1999, p.82-84).

Classification System ⇔ Description System

Food classification and food description may have very different goals, and this leads to very different appearances of the systems.

Food Classification Systems have been designed by and for people who know the foods involved and the uses that will be made of the data.

A description system is a tool of the data originator, who wants to give a description of the food, as precisely as possible, without the necessity of aggregating them.

DISCUSSION

Examples of classifying cheese

CIAA Food Categorization System – designed to serve as an allocation tool for food additives – first differentiates unripened, ripened, processed and analogue cheese; the second criterion is the conditioning, conservation and presence of rind.

In Eurocode 2 (food consumption surveys), cheeses are first classed depending on their consistency (hard, soft, fresh), then according to their fat content.

COICOP simply classes all cheeses under milk products group "cheese". The classifications are often contradictory, and their very existence shows that there can be no single satisfactory international classification system (Ireland J.D., 2000, p.533).

9.6.3. Purchases – food availability

HBS data refers to availability of food purchases entering the households and not the actual consumption of food intake of nutrients. Since average food purchases reflect the average food consumption the HBS food availability data reflect the consumption as well.

9.6.4. Larder stocks

The HBS method assumes that, within a given category of household composition, over a sufficient number of households, there is no change in the average levels of food stocks, although it is recognized that some households will acquire more food than they consume over the survey

DISCUSSION

period, while others will acquire less and use existing stocks to make up for any deficit.

9.6.5. Waste and visitors´ consumption of food

All edible waste includes scraping from plates, cooking and serving vessels, food that was spoiled, and food intended for human consumption but given to pets. Food eaten by visitors at meal times accounted on average for just fewer than 3% of the families´ "home" food consumption (Kanellou A., 1999, p.80-81).

9.7. NUTRIENTS

In addition to analyses of food availability, HBS have important potential for assessing nutrient intake levels. These should reflect nutrients as eaten rather than "purchased" and will depend on appropriate food consumption data which allow for changes in nutrient availability relating to food preparation and cooking.

Apart from the fact that HBS collect data on food quantities purchase and not necessarily consumed, another major problem is that in many cases, these data are given for food categories (e.g. seed oils) rather than single food items (e.g. olive oil) and different items are included under each food category. Additional information from various sources is needed, in order to decide on the proportion that different food items contribute to the nutrient values in a single food category (Trichopoulou A. and Lagiou P., 1999, p.25).

DISCUSSION

Assumptions must be made for the calculation of food consumption and nutrient supply, the consumption of different gender-, age-groups and regions cannot be differentiated considered. The consideration and comparison of the nutrient supply is only limited feasible (Schulze A., 1998, p.67).

Allowances for inedible material in the foods as purchased are also not included in the HBS. An edibility factor should be developed for calculating the edible portion. An essential element in converting the quantity data into nutrients is the development of appropriate conversion factors. These are based on food composition tables. However, in this connection a problem may be posed by the fact the list of food items in the HBS may not be in sufficient detail to enable a direct application of these food composition tables. This may be so in the case of minor foods, fruit or vegetables which tend to be grouped together. In such a case it is proposed to create certain "average conversion factors" on the basis of the mix of food items of particular food categories. This means that the lists of food items in the HBS need to be studied and a set of corresponding conversion factors has to be prepared (Trichopoulou A. and Lagiou P., 1997a, p.13-14).

Conversion of the food availability data to nutrients

The adequate diet could be obtained in many different ways and each of the types of food has a characteristic nutritional composition. Thus, research is needed to minimize the causes of variation between actual and measured diets.

DISCUSSION

Use of appropriate conversion factors may translate the food availability data into nutrients and allow estimation of nutrients consumed by an average. Hence, an essential element in converting the quantity data into nutrients is the development of appropriate conversion factors. These are based on food composition tables. However, the list of food items in the HBS may not be sufficiently detailed to enable a direct application of these food composition tables.

The conversion from food availability to nutrient intake, on a comparable European level, is a necessary purpose. It has to be assured that the countries are clear on the food definitions, which is achieved by a unified food classification, used and the limitations of their food composition tables, before converting the retrieved food availability data into nutrients (Kanellou A., 1999, p.85). For available food consumption data the use of the Euro Food Groups system was advised in order to make data on foods comparable in a European context.

Since national food composition tables do not allow comparisons of nutrient intakes between countries, it is necessary to obtain a standardization of European food composition data (Ireland J. et al., 2002, p.33).

It is not possible on an individual level, but on the group level to compare the nutrient supply with recommendations or guidelines because of the type of food record, due to the mode of coding of the food items in the food composition database (König J. and Elmadfa I., 1999, p.33).

Tasks aimed at developing methodologies for estimating the mean daily nutrient availability, using data from national HBS. The conversion of household food acquisitions into nutrient availability is not a

DISCUSSION

straightforward one, since HBS mostly provide data on the level of food groups, rather than on individual food items (DAFNE IV team, 2002a, p.7).

Participants of DAFNE IV acknowledged the necessity for a common food composition table to be used in a harmonized procedure for estimating nutrient availability from HBS data. Concerns were, however, expressed either in the use of the same composition data for all the countries or in the magnitude of errors that will be introduced if different national tables are used. Participants also concluded that the network should aim to a DAFNE Food Composition Table for DAFNE nutrient estimations (DAFNE IV team, 2002b, p.34).

9.8. FOOD EATEN OUTSIDE THE HOME

Changes in lifestyle factors – such as working patterns, family life, food shopping and storage patterns – have been a huge driver for changes in food consumption patterns in recent decades (Société française, 2000, p.14).

Eating patterns are changing in all the EU countries. First, people consume more meals away from home, especially youngsters, and secondly, eating patterns at home are also changing. As a result of changing socio-demographic characteristics, the number of meals eaten at home is decreasing, and conversely, increasing at restaurants, schools, and work-places. According to eating patterns at home, European food consumers can be split into two groups: northern-central European consumers, whose meals consist of a single dish, mainly meat, accompanied with vegetables, and southern European consumers whose meals are largely composed of multiple small dishes. Although this distinction is still valid, in the 1990s, eating patterns are changing in the countries (Gracia A. and Albisu L.M., 2001, p.479).

The lack of information on eating out is an important limitation of the HBS data and it is likely to affect estimations of food intake to a different extent in each country. The identification of dietary patterns however, is not expected to be seriously affected, since the type of food people choose to eat in their households is not remarkably different from the food they choose when they are eating out (Naska A. et al., 2005, p. 9).

The variety of foods and drinks eaten out causes some problems when estimating food and nutrient intakes. Estimated portion sizes and values may vary significantly for similar products (RimmerD.J., 2001, p.1174).

The only data concerning meals taken out-of-home that are collected in the HBS are expenses. The accessibility levels may have been underestimated as the data did not include meals bought and consumed away from the home, which may be significant to some households.

Efforts are made to estimate the proportion of the diet consumed from outside the household food supply, but usually no attempt is made to measure the consumption directly. It is generally assumed that the quality of the diet obtained away from is similar in nature to that consumed within the home (Kanellou A., 1999, p.82).

In practise it has been found that the accuracy of data of food consumed out of home obtained is doubtful. Some questions asked may be regarded as an interference into the privacy of the respondent and thus not answered

DISCUSSION

accurately. Further, some responses may be inaccurate as a result of poor memory. On the whole, such data can only be regarded as rough estimates. Methods of improving these data should be developed, and their inclusion in surveys should be encouraged so that the total food intake levels of the household population can be obtained (FAO, 1983, S.4).

In Austria 15-20% of the total meal expenditures are used for food and beverage acquisition outside home. The percentage cannot be specified exactly, because the expenditures in bars/restaurants do not allow a differentiation between food, drinks and other expenditures (Statistik Austria, 2003a, p. 7). Food and beverages consumed out of home were not recorded in quantities for the Austrian 1999 HBS.

A strong component of new product demand is for out-of-home food preparation and consumption. Pressed by time and a waning interest in food preparation, households increasingly demand frozen and chilled foods and pre-cut, pre-prepared meat, fruit and vegetables. These trends are also reflected in the growing percentage of household food expenditure spent on out-of-home food preparation, one of the most dynamic markets within the entire food chain. In Europe the food service industry is growing at 2-3% per year, against recent retail market growth of 5%.

In Austria the total household food expenditures of out-of home food sources is expected to reach 30 to 40%. More than a third of Austrians have their lunch outside the home, usually at work (OECD, 2001, p.15-16). This European trend can be explained by the fact of an increasing employment

DISCUSSION

of men and women, which leads to increased eating out, especially at fast-food restaurants (Leonhäuser I.-U. and Dorandt St., 2004, p.I/37).

Monthly food consumption expenditures of households, 1984 – 2000			
	1984	1993/94	1999/00
Share of food consumption expenditures in total household consumption	23,5 %	21,3 %	20,0 %
Thereof: share of expenditures of meals-out-of-home in the total food expenditures	21,4%	25,3 %	28,0 %

Figure 77. Food consumption expenditures of households in Austria (BLF, 2003, p.80)

9.8.1. DAFNE IV and meals out of home

All countries record information on expenses related to eating out occasions. In some countries (Finland and Portugal), some quantitive information is additionally collected, usually in relation to particular food items that can easily be assessed in pieces.

Expenses are usually grouped according to the place of consumption (e.g. expenses in restaurants; cafeterias; fast food outlets or take away).

Information on the type of meal consumed (breakfast, lunch, dinner, snack) is not always collected (information available in the Austrian dataset 1999/2000 and some data collected in the German survey).

A clear and common definition on what is meant by eating out is missing.

No information is available on the number of eating out occasions, within one day.

DISCUSSION

In case of meals, no information is collected on the number of people sharing the meal.

In most European HBS datasets, the main diary keeper is also for monitoring the out of home food expenses of all household members. In Greece and Portugal, each member above a certain age (e.g. 14 years old in Greece) is given a separate questionnaire.

Data are collected on the basis of a pre-defined list of items or a schedule of the acquisitions of participants-this data are aggregated and coded at a later stage (DAFNE IV team, 2002c, pp.215).

DISCUSSION
9.9. AVERAGE DAILY AVAILABILITY IN THE DAFNE COUNTRIES, IN THE 1990s

Food disparities in Austria and in the other European DAFNE member countries are documented in the following graphs. For demonstration, the most interesting food patterns are presented.

Data retrieved from the DAFNE databank

The average fruit availability is relatively high in Mediterranean countries, but is also considerable in Austria. The distribution of fruit varieties differs in the European countries, for instance Austria presents the highest and Ireland the lowest daily availability of apples (respectively 64 and 17 g/p/d).

Average availability of total meat and meat products in the 1990s

Disparities are observed in the type of meat preferred in the different European regions. Processed meat is particularly consumed in Central Europe, for example 94 g/p/d in Austria, but only 9 g/p/d in Greece. The availability of pork varies from 55 g/p/d in Hungary to 10 g/p/d in Ireland.

DISCUSSION

Table 24. Average daily availability of fresh fruits by type in the DAFNE countries, in the 1990s (g/p/d)

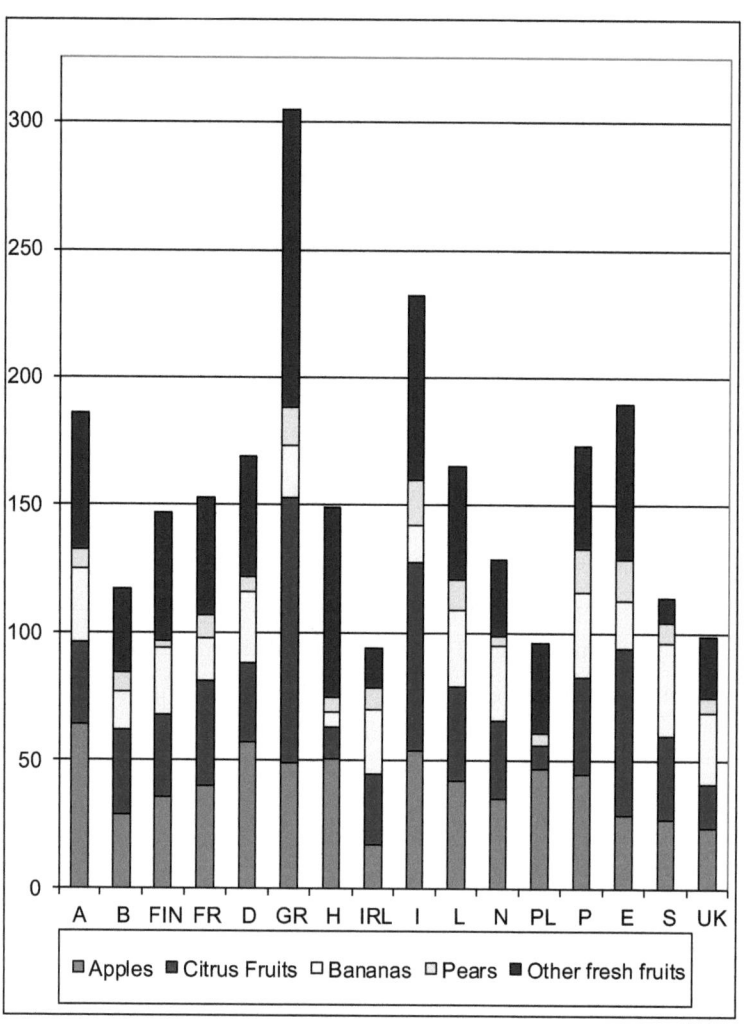

DISCUSSION

Table 25. Average availability of total meat and meat products in the 1990 (g/p)

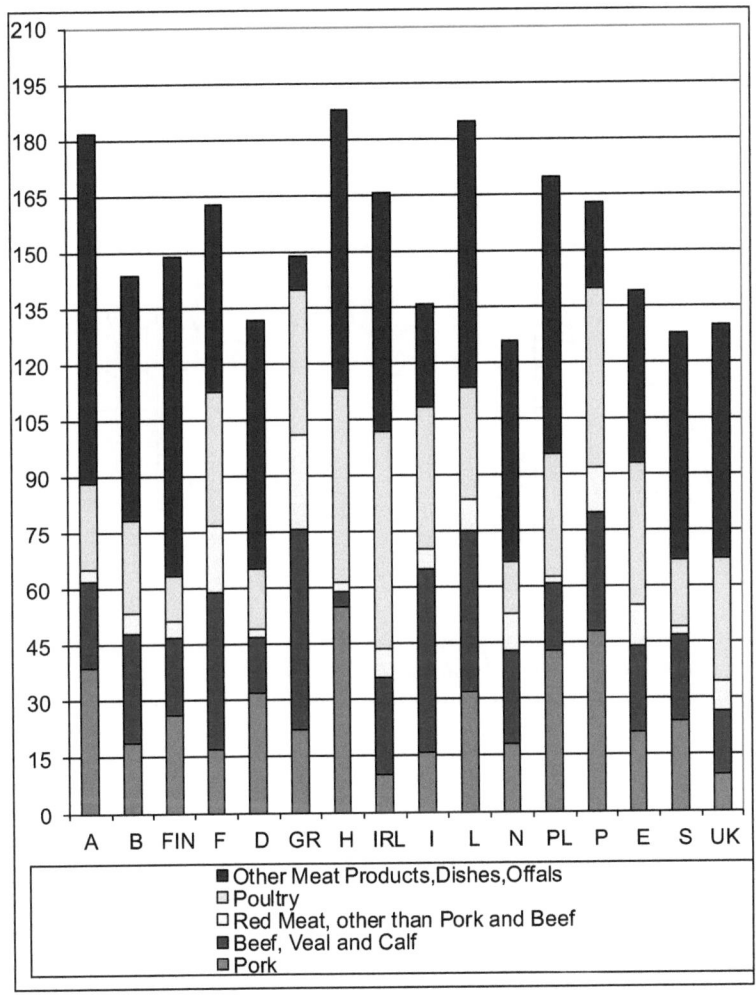

DISCUSSION

Table 26. Average daily availability of vegetables in the DAFNE countries, in the 1990s g/p/d)

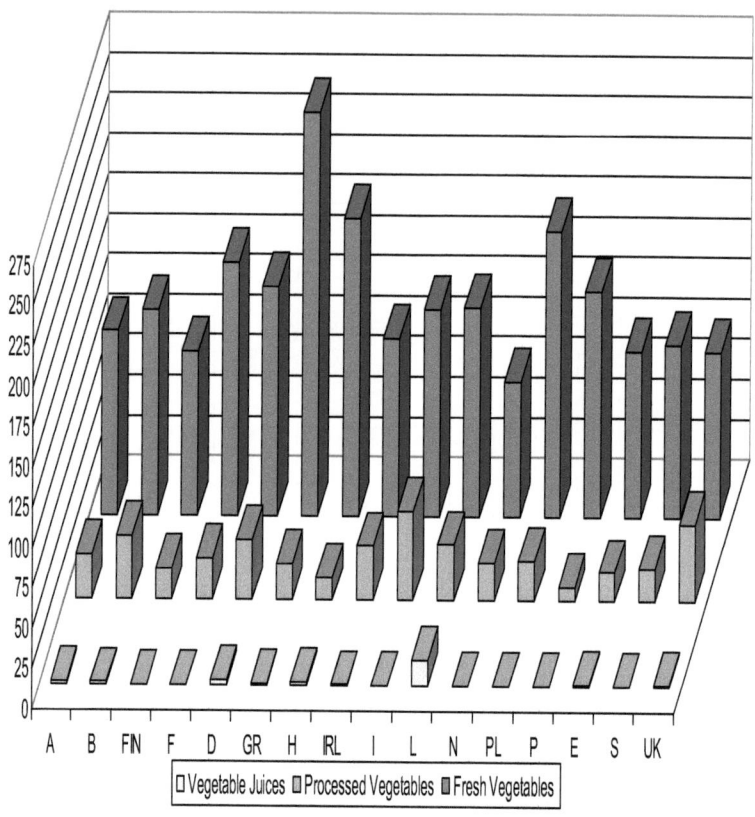

Countries which are high vegetable consumers reported the lowest availability of vegetable juices.

Greece shows the highest, whereas Norway has the lowest availability of fresh vegetables (251 and 84 g/p/d, respectively).

DISCUSSION

The Mediterranean country Italy recorded the highest availability, whereas another Mediterranean country, namely Portugal recorded the lowest availability (56 and 8,72 g/p/d, respectively).

Table 27. Daily Availability of cheese in the DAFNE countries in the 1990s (g/p/d)

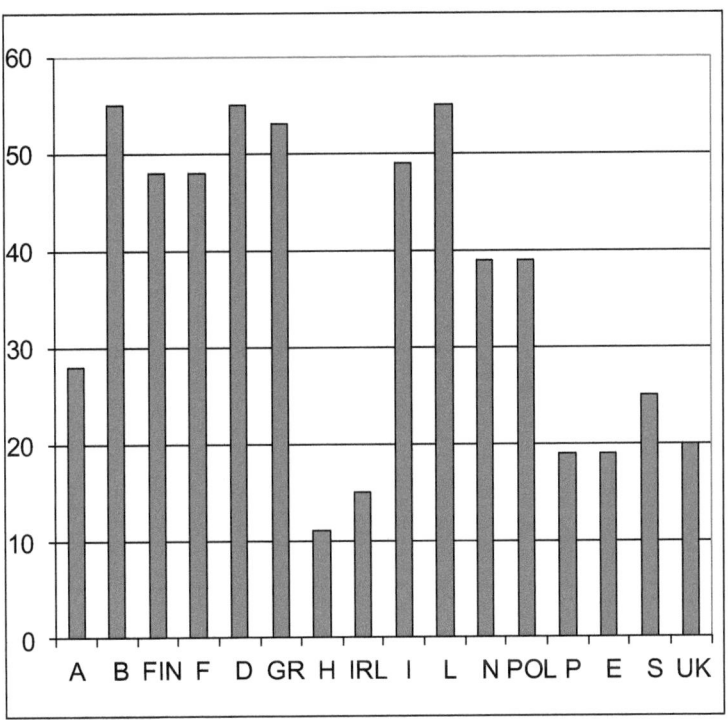

In Belgium, Germany, Greece and Luxembourg are the high consumers of cheese; for instance 55 g/p/d in Belgium, Germany, Luxembourg and 53 g/p/d in Greece, whereas only 11 g/p/d in Hungary.

DISCUSSION

Table 28. Daily availability of milk in the DAFNE countries in the 1990s (ml/p/d)

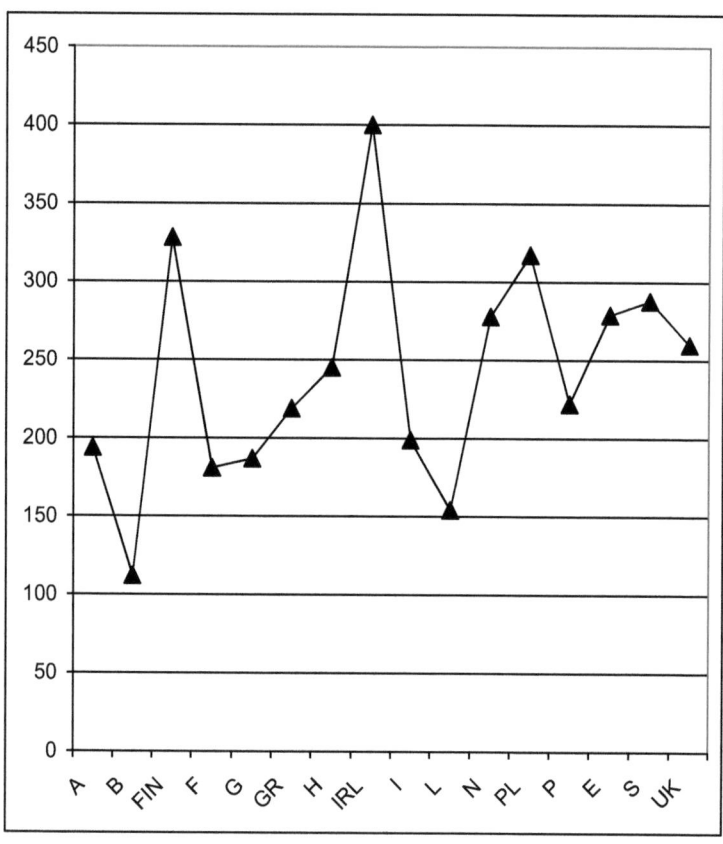

In Europe the milk availability ranges from 400 g/p/d in Ireland to 112 g/p/d in Belgium.

DISCUSSION

Table 29. Daily availability of cheese and milk products in the DAFNE countries in the 1990s (g/p/d)

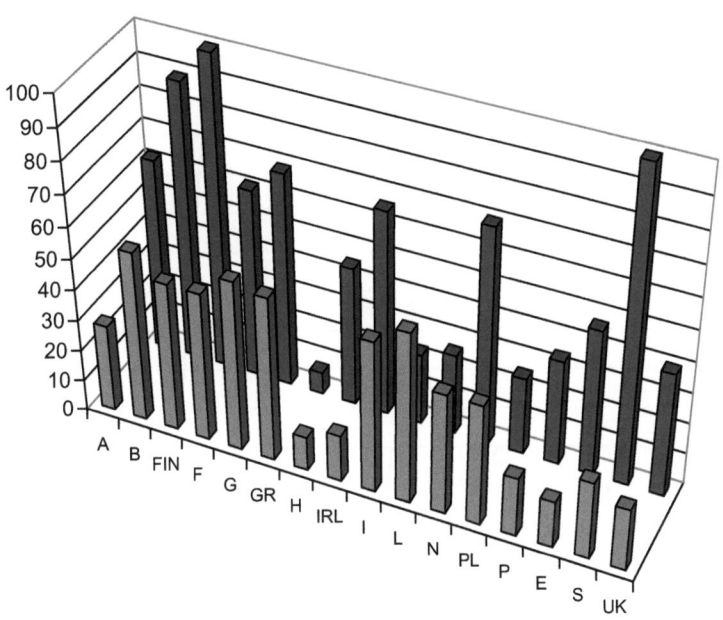

DISCUSSION

Greece is a higher consumer of cheese availability, but shows the lowest availability of milk products. Finland and Sweden recorded the highest availability of milk products.

Table 30. Daily availability of sugar and sugar products in the DAFNE countries in the 1990s (g/p/d)

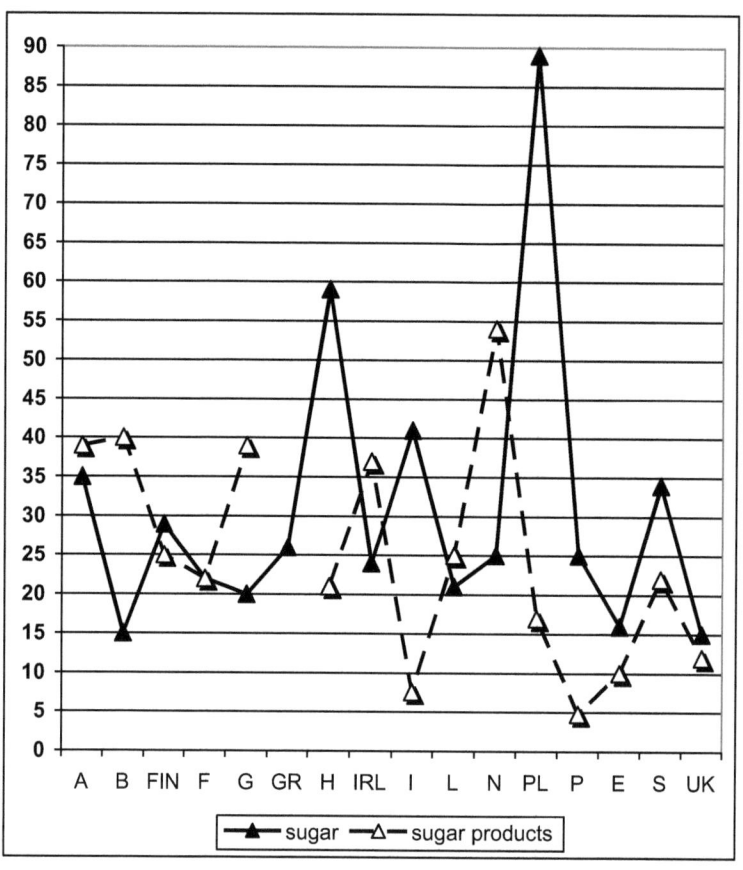

DISCUSSION

The daily availability of sugar ranges from 15 g/p/d in Belgium and the UK to 89 g/p/d 89 g/p/d in Poland. The availability of sugar products ranges from 4,75 g/p/d in Portugal to 54 g/p/d in Norway.

Table 31. Daily availability of potatoes in the DAFNE countries in the 1990s (g/p/d)

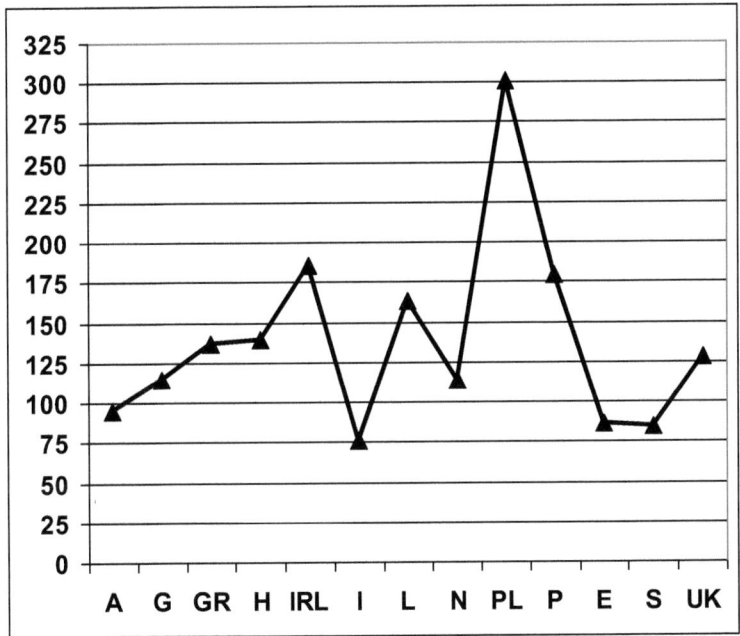

In Europe the availability of potatoes ranges from 76 g/p/d in Italy to 301 g/p/d in Poland.

DISCUSSION

Table 32. Daily availability of nuts in the DAFNE countries in the 1990s (g/p/d)

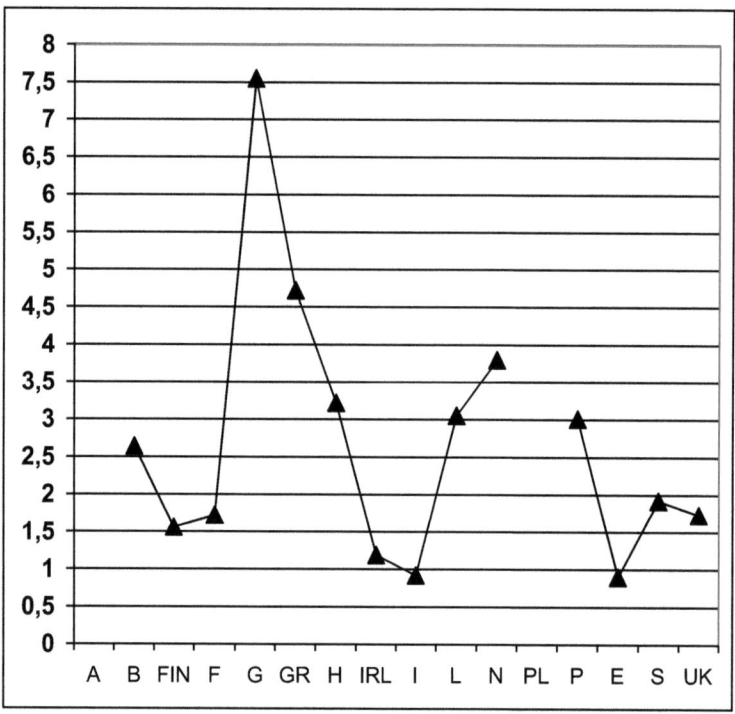

Germany shows the highest, whereas Spain has the lowest daily availability of nuts (7,55 and 0,9 g/p/d, respectively). In Austria and Poland separated values of nuts are not available.

DISCUSSION

Table 33. Daily availability of pulses by mean in the DAFNE countries in the 1990s(g/p/d)

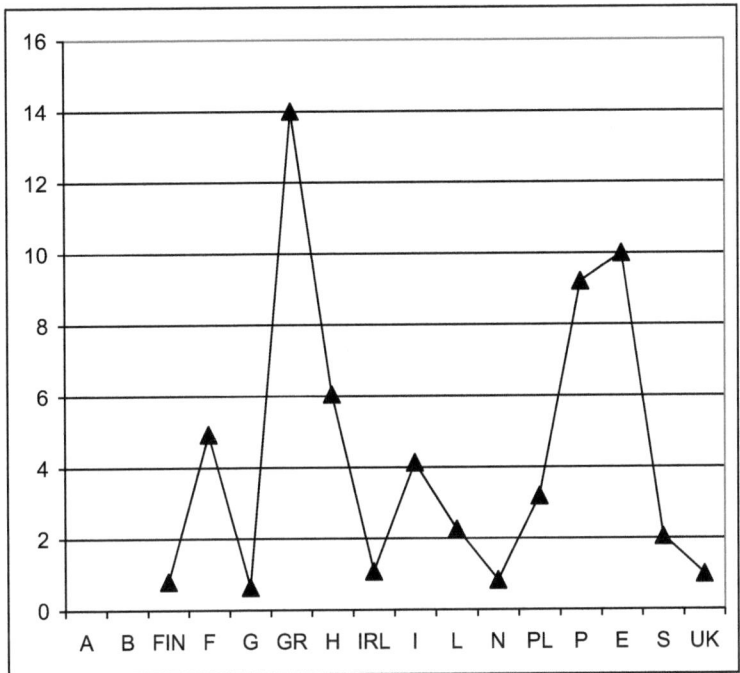

The Mediterranean countries Greece, Spain and Portugal are the higher consumers of pulses. The Northern European countries Finland, Ireland, Norway, the UK and the Central European country are lower consumers. In Austria and Belgium separated availability values of pulses are not available.

10 CONCLUSIONS AND RECOMMENDATIONS

10.1. AUSTRIAN TRENDS IN DAILY FOOD AVAILABILITY

Due to the fact that previous HBS are not comparable with the results of HBS 1999/2000 as a result of differences in methodology and standards, Austrian trends in daily food availability based on HBS cannot be described, but incorporation of new data in the database will provide for this possibility.

Especially this concerns the application of the nomenclature COICOP, which because of conceptual differences in comparison to the applied national systematics of HBS 1993/94.

The international comparability of the Austrian HBS must be improved in comparison to the previous HBS. This can be traced to the Austrian membership of the European Union and the enforcement of European standards in the survey.

10.2. ASSOCIATION OF AUSTRIAN FOOD AVAILABILITY WITH SOCIOECONOMIC CHARACTERISTICS

HBS provide data about food availability of persons in association with their socioeconomic characteristics.

Food availability by locality shows that rural households have the highest availability of sugar and products, potatoes and food of animal origin except the lowest value for fish/seafood. The semi-urban availability values are the highest for alcoholic beverages and fruits, and the lowest for milk

CONCLUSION AND RECOMMENDATIONS

and products. In urban areas households have the lowest availability of alcoholic beverages, but the highest availability of non alcoholic beverages, fruit and vegetable juices.

Households headed by retired persons have the highest availability of all food groups except of fruit and vegetable juices which is the lowest value of all occupation categories.

Elderly single-households have more food availability of all food groups with the exception of alcoholic, non-alcoholic beverages and fruit and vegetable juices than adult single-households.

"Unemployed" households have higher daily food availability of cereals, milk and milk products, meat, vegetables, fish and seafood, fruits, added lipids, alcoholic beverages, fruit and vegetable juices than "manual" and "non-manual" households. Except: For eggs, non alcoholic beverages, sugar and sugar products, "manual" households show higher values.

"Elderly" and "retired" households have more food available at household level compared with the other "household composition" and "occupation" groups. Hence, it is likely that elderly people are eating at home in most instances. It can be expected that employed people have meals away from home during their breaks. The missing information about meals consumed out of home does not affect employed and unemployed households equally.

Across the four socio-economic DAFNE-categories "adults+children", "urban", "non-manual" and "higher educated" households show the lowest availability of eggs and added lipids.

The daily food availability of potatoes, eggs, meat, added lipids, non alcoholic beverages, sugar and sugar products tends to decrease with

CONCLUSION AND RECOMMENDATIONS

educational level. The availability of fish and seafood increases with educational level. The daily availability of fish by locality of the household shows the same increasing pattern with the urbanization grade.

The average availability of eggs, potatoes and meat decreases with increasing degree of urbanization and educational level. The availability of lipids, non alcoholic beverages and sugar and sugar products tends to decline with educational level, whereas the availability of cereals, vegetables, non alcoholic beverages, fruit and vegetable juices tends to increase with the degree of urbanization.

10.3. DISPARITIES IN FOOD PATTERNS IN EUROPE

Food consumption varies in Europe; to get a deeper understanding of the variation it would be useful to gather more comparable information from countries representing the different regions. It would be important to obtain a better understanding of how diet contributes to the social differentials in health and whether socio-economic status modifies associations between diet and disease. A prerequisite for analysing the role of diet in inequalities of health is to have an understanding of the role of health behaviours. Therefore, there it is also a need for studies on socio-economic differences in food habits in relation to other health behaviours and lifestyle. The differences in the patterns of disparities between regions need to be considered when effort to improve nutrition and health are planned. In northern Europe it could, for example, be effective to address the question of how to direct the increase in vegetable consumption more to those with low education. In the south, the traditional diet includes vegetables and it is

CONCLUSION AND RECOMMENDATIONS

therefore relevant to try to keep the traditional diet includes vegetables and prevent the low socio-economic groups from adopting "northern" habits (Roos G., 2001, p.36-40). Differences between classes in terms of modern and traditional, it is important to remember that age can exert an even stronger influence (Rowely R., 2006, p.15).

Europe's Member States bear heavy economic, public and social costs because of diet-related diseases, poor nutrition and a sedentary lifestyle. Conversely, healthy eating and physical activity can lead to a healthy society and promote a healthy economy.

The nutritional status of individuals, which influences in a major way their current and future state of health, is the result of a balance between their nutritional needs and the supply of nutrients available to them (Société française, 2000, p.12-14).

CONCLUSION AND RECOMMENDATIONS

10.4. AUSTRIAN RESULTS IN COMPARISON TO THOSE OF THE OTHER DAFNE-COUNTRIES IN THE NINETIES

In Austria and Germany households recorded a **fruit** consumption of more than 180g/person/day.

Hungary (80g/person/day) and Norway (79 g/person/day) have higher daily availability of **sugar and sugar products** than Austria (74g/person/day). The United Kingdom (31g/person/day), Portugal (34g/person/day) and Spain (35g/person/day) recorded lower daily availability of **sugar and sugar products**.

Hungary recorded the lowest **fish and seafood** availability (4.4 g/person/day), Austria recorded 9.3 g/person/day in 1999/2000. Portugal is leading **fish** availability with 85g/person/day, followed by Spain (74 g/person/day), Norway (50 g/person/day) and Greece (45 g/person/day).

Greece recorded the highest **vegetable** availability within the household (271 g/person/day); Norway has the lowest **vegetable** availability (109 g/person/day). Austria (142g/person/day) has the lowest **vegetable** availability of all Central European countries.

The highest availability values of **alcoholic beverages** were recorded in Germany (200 ml/person/day) and the lowest in Greece (48 ml/person/day) and the UK (51 ml/person/day). Austria has a daily **alcoholic beverages** availability of 171 g/person/day.

In comparison to Germany Austria recorded lower values of **non alcoholic beverages** (915ml/person/day and 652 ml/person/day, respectively).

Hungary and Austria recorded the highest **meat and meat products** availability (190g/person/day and 182g/person/day, respectively), Norway the lowest availability (126g/person/day).

10.5. INFORMATION OF FOOD AVAILABILITY COLLECTED BY HBS, ADVANTAGES AND DISADVANTAGES

Data comparability is important to find out the differences and similarities in nutrition habits and the health status of European population (Elmadfa I. and Weichselbaum E., 2005b, p.63), this has important implications for the design of future food consumption surveys (Flynn A., 2001, p.1127).

The fact is that despite their limitations, HBS generate nutrition data at regular time intervals in all European countries. This information provides a valuable background for the conduct of a wide range of nutrition analyses in Northern, Central and Southern European countries. It also constitutes an affordable alternative to most North, Central and South European countries. HBS data could help identify issues such as differences concerning dietary patterns, high risk population groups on account of their nutritional habits, relationships between diet, morbidity and mortality data, and the dietary intake of additives and contaminants.

Nutrition data from HBS represent an important source of nutrition information. The data are useful for international comparisons, as well as for a reliable assessment of time trends. Utilization of HBS data could be of great importance for the realization of various purpose including nutrition and agricultural strategy planning and marketing purposes of food

CONCLUSION AND RECOMMENDATIONS

industries. Food and agricultural policy strives to provide stable, nutritional and affordable food supplies (Kanellou A., 1999, p.89-90).

FAO, HBS and individual dietary survey data suggest that there remains considerable room for improvement in Europe's diets and also differences in dietary patterns among European States. Recent dietary survey data suggest that very few countries are consuming more than the daily of 400g of fruit and vegetables recommended by the World Health Organization (Société française, 2000, p.6).

The DAFNE project purposes the monitoring of nutritional patterns and produces results from European countries that are comparable in terms of definitions of food groups, occupation, household composition and locality (Paterakis S.E. and Nelson M., 2003, p.579).

The potential of the DAFNE network is the easy access to reliable and comparable data in depicting dietary trends, consumer behaviour and nutritional habits (Trichopoulou A., 2005, p.2).

System providing routine food and nutrition surveillance in local and national populations are essential to the understanding of relations between nutrition and health. An integrated HBS food database and common standards in the methods used in different European countries for the HBS food data collection allow the identification of differences concerning dietary patterns and high risk population groups owing to their nutritional habits.

The collection of food consumption at a household level can be more economical than individual surveys for obtaining information for ecological studies.

CONCLUSION AND RECOMMENDATIONS

High quality data on food intake are necessary to study relationships between diet and health. This information may help to further design nutrition policy programs at the European level but also at regional or national levels (Kanellou A., 1999, p.88).

HBS have several advantages (Trichopoulou A., 1992, p.4): they are conducted regularly in most European countries with a time interval which varies from 1 to 7 years; they use representative samples of households; they generate a substantial amount of data concerning nutrition; they allow cross-linkage to socio-demographic characteristics of the households, which could be useful for standardization and exploratory analyses.

HBS have several limitations. HBS nutrition data are different from one country to another, not only in relation to the number of food items which are recorded but also the type of information provided. No information about the consumption of food commodities and beverages outside the household are recorded, although information about the expenditure involved in these meals is available.

In most countries no information is collected concerning losses and waste of food, but the resulting overestimation of food availability for the human population groups.

In addition, estimation of nutrient intakes from HBS food availability data requires that a series of assumptions and approximations be made, because most countries collect data only for large food groups. Finally, there is no uniform coding system, and rules must be developed and agreed for the aggregation of food items appearing in the HBS (Trichopoulou A. et al., 1996, p.700).

CONCLUSION AND RECOMMENDATIONS

The HBS food data fail in two major areas: completeness of data and knowledge of food and nutrient within households. A HBS nutrition database is clearly imperfect, but HBS may represent the best of existing realistic alternatives. It can provide information at least as good, and probably better, than that obtained through the FBS (Kanellou A., 1999, p.95).

Despite these limitations the analyses show that household purchase data can be a valuable tool for obtaining information on the food pattern of a population (Becker W., 2001, p.1181).

In a world dominated by rapid changes, often dictated by the principles of globalisation, the ability to assess, monitor and compare dietary patterns of different populations is of paramount importance. The DAFNE refers to collaborative effort of participants around Europe to develop a bank of regularly updated and comparable dietary data, based on HBS (Trichopoulou A. and Naska A., 2001, p.1127) and improves the understanding of factors that affect food choices and provide a link between an individual's knowledge and attitudes and his or her dietary behaviour (Tippett K.S., 1999, p.6).

Careful use of HBS data is necessary to avoid falling to potential traps of misinterpretation. Several aspects of the HBS results need further investigation, such as the ratio of food eaten away from home to total consumption, similar food group questionnaires and food coding systems for all countries.

11   SUMMARY

If it is not available, it will not be eaten
If it is available, it is likely to be eaten
If there is no alternative, it will be eaten (Mela D.J., 1999, p.514)

The aim of this thesis was to compile the required DAFNE tasks for DAFNE IV – participant Austria; to undertake the task of classifying the food information collected in the Austrian HBS according to the agreed DAFNE food classification system and to group the four socio-demographic DAFNE variables according to the DAFNE classification schemes. Furthermore, this thesis analyses in regard to the Austrian DAFNE IV results the Austrian food consumption by socio – economic and demographic parameters, to compare these data with data of other DAFNE members and to consider the advantages and disadvantages of data derived from HBS.

This thesis demonstrates that it is possible to integrate the Austrian HBS data into the `Data Food Networking´ nutrition databank. Bearing in mind the advantages and disadvantages of HBS, the DAFNE databank constitutes a cost – effective European databank, based on food, socio – economic and demographic data, which has the potential to serve for the nutritional and agricultural strategy planning at regional, national and international level.

Food availability certainly does vary across Europe: over 600g/p/d fruit and vegetables are available daily to the consumers of Greece compared to 334g/p/d of Austrians and 275g/p/d of Norwegians consumers, according

SUMMARY

to data from DAFNE. Within countries there are also considerable discrepancies in the availability of some foods, such as fresh fruit and vegetables.

Except from the overall means, the Austrian food availability by socio-demographic parameters has shown interesting trends, for instance that the higher educated and the urban households behave often similar in regard to their food consumption.

Austrian "elderly" and "retired" households have more food available at household level compared with the other socio-demographic groups, since it is likely that elderly people eat at home and employed people rather have meals away from home.

Germany (200 ml/p/day) and Austria (171 ml/p/day) show higher values of alcoholic beverages at household level than Greece (48 ml/p/day) and the UK (51 ml/p/day).

Information about food losses, food and beverages consumed outside home and in institutional households and the consideration of age and sex are missing. But since it is unrealistic to contemplate an international system for monitoring intakes at the individual level, HBS may represent the best of existing realistic alternatives. HBS nutrition databases are clearly imperfect, but provide probably better nutrition information than FBS of the Food and Agriculture Organization.

Hence, the multipurpose HBS are routinely undertaken by the National Statistical Offices of almost all European countries with a methodology which allows for between countries comparisons, the HBS data could help identify national and international disparities concerning dietary patterns.

ZUSAMMENFASSUNG

Das Ziel dieser Arbeit war es die geforderten DAFNE Aufgaben für den DAFNE IV – Teilnehmer Österreich zu erfüllen; die Aufgabe der Klassifizierung, der in den österreichischen HBS gesammelten Ernährungsinformationen, nach dem festgesetzten DAFNE Lebensmittelklassifizierungssystem und den vier soziodemographischen DAFNE Variablen. Weiters analysiert diese Arbeit hinsichtlich der österreichischen DAFNE IV Ergebnisse den österreichischen Lebensmittelverbrauch nach sozioökonomischen und –demographischen Parametern, diese Daten werden mit Daten anderer DAFNE Mitglieder verglichen und die Vor- und Nachteile der von HBS abgeleiteten Daten betrachtet.

Diese Arbeit demonstriert, dass die österreichschen HBS Daten für die Integration in die `Data Food Networking´ Ernährungsdatenbank geeignet sind. Unter Berücksichtigung der Vor- und Nachteile der HBS, stellt die DAFNE Datenbank eine kosteneffektive Europäische Datenbank dar, basierend auf Lebensmittel, sozioökonomischen und –demographischen Daten, welche das Potential der Ernährungs- und agrarwirtschaftlichen Strategieplanung auf regionalem, nationalem und internationalem Niveau hat.

Zweifellos variiert die Lebensmittelverfügbarkeit quer durch Europa: über 600g/p/d Obst und Gemüse sind täglich den Konsumenten in Griechenland verfügbar, im Vergleich dazu 334g/p/d den österreichischen und 275g/p/d den norwegischen Konsumenten, gemäß den DAFNE Daten. Innerhalb der

ZUSAMMENFASSUNG

Länder gibt es auch beachtliche Abweichungen in der Verfügbarkeit einiger Lebensmittel, wie zu Beispiel frisches Obst und Gemüse.

Abgesehen von den allgemeinen Durchschnittswerten, zeigt die österreichische Nahrungsmittelverfügbarkeit anhand der soziodemographischen Parameter interessante Trends, zum Beispiel, dass höher gebildete und städtische Haushalte sich oft in Bezug auf ihren Lebensmittelverbrauch ähnlich verhalten. Österreichische „ältere" und „pensionierte" Haushalte verbrauchen auf Haushaltsebene mehr Lebensmittel als andere soziodemographischen Gruppen, da wahrscheinlich ältere Menschen eher zu Hause essen und beschäftigte Menschen eher Speisen außer Haus konsumieren.

Deutschland (200 ml/p/Tag) und Österreich (171 ml/p/Tag) zeigen höhere Werte alkoholischer Getränke auf Haushaltsniveau als Griechenland (48 ml/p/Tag) und das UK (51 ml/p/Tag).

Informationen über Lebensmittelverluste, Nahrungsmittel und Getränke außer Haus konsumiert, Informationen über institutionaler Haushalte und die Berücksichtigung des Alters und des Geschlechts fehlen.

Da die Erwägung eines internationalen Systems auf individuellem Niveau unrealistisch ist, stellen HBS die beste existierende realistische Alternative dar.

Die Mehrzweck HBS werden routinemäßig von den Statistikämtern in geradezu allen europäischen Ländern mit einer Methodologie, die zwischenstaatliche Vergleiche berücksichtigt, durchgeführt; die HBS Daten können die Identifizierung nationaler und internationaler Ungleichheit bezüglich Ernährungsmuster unterstützen.

12 REFERENCES

Australian Institute (1996) of Health and Welfare: Australia's Health 1996 – Improving the health of Australians, p. 1-43 www.aihw.gov.au/publications/health/ah96/ah96-c03.html, (14/12/2006)

Becker W. (2001): Comparability of household and individual food consumption data- evidence from Sweden, in: Public Health Nutrition 4(5B), 2001, p.1177-1182

Becker W. and Helsing E. (1993): Ernährungs- und Gesundheitsdaten – Ihr Nutzen für eine Ernährungspolitik, Kopenhagen: Regionale Veröffentlichung der Weltgesundheitsorganisation, Regionalbüro für Europa, Europäische Schriftenreihe Nr.34, 1993, p.1-192

Bellisle F. (1999): Food choice. Appetite and physical activity, in: Public Health Nutrition, Volume2: Issue 3(A), Supplement: 1, 1999, p.357-361

Biro G., Hulshof KFAM., Ovesen L. (2002): Selection of methodology to assess food intake, in: European Journal of Clinical Nutrition 56, Suppl 2, 2002, p.S25-S32

BLF (2003): 2. Lebensmittelbericht Österreich– Die Entwicklung des Lebensmittelsektors von 1995 bis 2002, Bundesministerium für Land- und Forstwirtschaft, Umwelt und Wasser, 2003, p. 1-139

Blisard N., Stewart H., Jolliffe D. (2004): Low-Income Households´ Expenditures on Fruits and Vegetables, in: USDA Agricultural Economic Report Number 833, 2004, p.1-27

Blundell J.E. (2000): What foods do people habitually eat? A dilemma for nutrition, an enigma for psychology, in: American Journal of Clinical Nutrition, Vol. 71, No. 1, 2000, p.3-5

Breslin L. (2001): Foreword, in: Public Health Nutrition 4(5B), 2001, p.1129

REFERENCES

Byrd-Bredbenner C., Lagiou P., Trichopoulou A. (2000): A comparison of household food availability in 11 countries, in: Journal of Human Nutrition Dietetics, 13, 2000, p. 197-204

Dafne data bank: DAFNE software –
http://www.nut.uoa.gr/DafneSoftWeb/ (18.02.2007)

DAFNE III team (1999): European Food Availability databank based on Household Budget Surveys: – Executive Summary Report of the DAFNE III project, Grant Agreement number S12.195600, 1999, p. 5-7
http://ec.europa.eu/comm/health/ph_projects/1999//monitoring/fp_monitoring_1999_exs_01_en.pdf (14/12/2006)

DAFNE IV team (2002a): European Food Availability databank based on Household Budget Surveys: – Executive Summary Report of the DAFNE IV project, Grant Agreement number SPC.2002336, 2002, p. 1-15
http://ec.europa.eu/comm/health/ph_projects/2002/monitoring/fp_monitoring_2002_exs_04_en.pdf (14/12/2006)

DAFNE IV team (2002b): European Food Availability databank based on Household Budget Surveys: Final Activity Report of the DAFNE IV project, Grant Agreement number SPC.2002336, 2002, p. 1-53
http://ec.europa.eu/comm/health/ph_projects/2002/monitoring/fp_monitoring_2002_frep_04_en.pdf (14/12/2006)

DAFNE IV team (2002c), Protocol for collecting information on eating out in the HBSs, Annex V, 2002, p.215-219
http://ec.europa.eu/comm/health/ph_projects/2002/monitoring/fp_monitoring_2002_annexe5_04en.pdf (14/12/2006)

De Irala-Estevez J., Groth M., Johansson L. (2000): A systematic review of socio-economic differences in food habits in Europe: consumption of fruit and vegetables, in: European Journal of Clinical Nutrition, 54, 2000, p.706-714

De Jong N. (1999): Conference report third international dietary assessment methods congress: Implications for future policy, in: Nutrition Today, May, 1999, p. 1-12

REFERENCES

Deutsche Gesellschaft für Ernährung, DGE (2004): DGE-Info Wissenschaft 04/2004, Der neue DGE-Ernährungskreis

Dofkova M., Kopriva V., Resova D. (2001): The development of food consumption in the Czech Republic after 1989, in: Public Health Nutrition: 4(5), 2001, p.999-1003

Elmadfa I. and Weichselbaum E. (2005a): European Nutrition and Health Report 2004, Forum Nutr. Basel, Karger, vol.58, 2005, p.1-224

Elmadfa I. and Weichselbaum E. (2005b): On the nutrition and health situation in the European Union, in: Journal of Public Health, 13, 2005, p.62-68

Elmadfa I. and Suchomel A. (2002): Trends in food availability in Austria – the DAFNE IV project, 2002, p. 1-20
http://ec.eu/comm/health/ph_projects/2002/monitoring/fp_monitoring_200 2_annexe_aust_04_en.pdf (14/12/2006)

EFCOSUM group (2001): European Food Consumption Survey Method – Final Report, TNO report, 2001, p.1-79

The EFCOSUM Group (2002): Summary – European Food Consumption Survey Method, in: European Journal of Clinical Nutrition 56, Suppl 2, 2002, p. S1-S3

EU Commission (2005): The DAFNE Food Classification System Operationalisation in 16 European, 2005, p. 1-83

Eurostat (2003): European Commission, HBS in the EU-Methodology and recommendations for harmonisation, Methods and nomenclature, Office for Official Publications of the European Communities – Luxembourg, 2003, p.1-53

Eurostat (2007) / U.S. Bureu of the Census, http:///.eurostat.ec.europa.eu (20 April 2007)

FAO Statistic Division (1983): FAO Economic and Social Development Paper 34 – A comparative study of food consumption data from food

REFERENCES

balance sheets and household surveys, Food and Agriculture Organization of the United Nations, Rome, 1983, p. 2-89

Flynn A. (2001): Conclusions – The North/South Ireland Food Consumption Survey, in: Public Health Nutrition 4(5A), 2001, p.1127

Friel S., Nelson M., McCormack K., Kelleher C. (2001): Methodological issues using household budget survey expenditure data for individual food availability estimation: Irish experience in the DAFNE pan-European project, in: Public Health Nutrition 4(5B), 2001, p.1143-1147

Galobardes B., Morabia A., Bernstein M.S. (2001): Diet and socioeconomic position: does the use of different indicators matter?, in: International Journal of Epidemiology; 30, 2001, p.334-340

Gedrich K. (1996): Ökonometrische Bestimmung der Lebensmittel- und Nährstoffzufuhr von Personen anhand des Lebensmittelverbrauchs von Haushalten, Studien zur Haushaltsökonomie 13, PETER LANG, 1996, p. 7-226

Gil Jose M. (1995): Food consumption and economic development in the European Union, in: European review of agricultural economies 22/3, mouton de gruyter, 1995, p. 385-400

Gil Jose M., Gracia A., Angulo A.M. (2000): Trends in the consumption of animal food products in Mediterranean countries, CIHEAM – Options Mediterraneenes, 2000, p.11-31

Gracia A. and Albisu L.M. (2001): Food consumption in the European Union: Main determinants and country differences, in: Agribusiness, Volume 17, Issue 4, 2001, p.469-488

Harro M., Villa I., Liiv Kr. (2005): Nutrition-related health indicators and their major determinants in the new member state: case of Estonia, in: Journal of Public Health, Volume 13, Number 2, 2005, p. 111-119

Ireland J.D. (2000): Review of International Food Classification and Description – Critical Review, Journal of Food Composition and Analysis, 13, 2000, p. 529-538

Ireland J., van Erp-Baart AMJ., Charrondiére UR. (2002): Selection of a food classification system and a food composition database for future food consumption surveys, in: European Journal of Clinical Nutrition 56, Suppl 2, 2002, p. S33-S45

James W.Ph., Nelson M., Ralph A. (1997): Socioeconomic determinants of health: The contribution of nutrition to inequalities in health, in: British Medical Journal; 314, 1997, p. 1545-1553

Kanellou A. (1999): Methodological approach for the harmonization of household budget surveys' food data – a comparison among the German, Greek and Hungarian population, Ernährungswissenschaften, Fachverlag Köhler Giessen, 1999, p. 1-72

Kelleher C., Friel S., Nolan G. (2002): Symposium on „Influence of social and cultural variations diet" – Effect of social variation on the Irish diet, in: Proceedings of the Nutrition Society, 61, 2002, p. 527-536

König J. and Elmadfa I. (1999): Food-based dietary guidelines – the Austrian perspective, in: British Journal of Nutrition, 81, Suppl.2. 1999, p. 31-35

Kroes R., Müller D., Lambe J. (2002): Assessment of intake from the diet, Food and Chemical Toxicology 40, 2002, p. 327-385

Lagiou P. and Trichopoulou A. (1999): Household budget survey nutritional data in relation to mortality from coronary heart disease, colorectal cancer and female breast cancer in European countries, in: European Journal of Clinical Nutrition, 53, 1999, p. 328-332

Lagiou P. and Trichopoulou A. (2001): The DAFNE initiative: the methodology for assessing dietary patterns across Europe using household budget survey data, in: Public Health Nutrition 4(5B), 2001, p.1135-1141

Leonhäuser I.-U. and Dorandt St. (2004): The benefit of the Mediterranean diet – Considerations to modify German food patterns, in: European Journal of Nutrition Volume 43, Supplement 1, 2004, p.i31-i38

REFERENCES

Lindström M., Hanson B.S., Wirfält E. (2001): Socioeconomic differences in the consumption of vegetables, fruit and fruit juices – The influence of psychosocial factors, in: European Journal of Public Health; 11, 2001, p.51-59

Marshall D. (1995): Food Choice and the Consumer, Blackie Academic & Professional, 1995, p. 1-274

Mela D.J. (1999): Symposium on "Functionality of nutrients and behaviour" – Food choice and Intake: the human factor, in: Proceedings of the Nutrition Society, 58, 1999, p.513-521

McKee M. and Ryan J. (2003): Monitoring health in Europe: opportunities, challenges, and progress, in: European Journal of Public Health; 13 (3 Supplement), 2003, p.1-4

Meiselman H.L. (1996): Food Choice, Acceptance and Consumption, Blackie Academic & Professional, 1996, p. 1-383

Naska A., Vasdekis VGS., Trichopoulou A. (2000): Fruit and vegetable availability among ten European countries: how does it compare with the "five-a-day" recommendation?, in: British Journal of Nutrition, 84, 2000, p.549-556

Naska A., Paterakis S., Beckman H. (2001a): Methodology for rendering household budget and individual nutrition surveys comparable, at the level of the dietary information collected, in: Public Health Nutrition: 4(5B), 2001, p.1153-1158

Naska A., Vasdekis VGS., Trichopoulou A. (2001b): A preliminary assessment of the use of household budget survey data for the prediction of individual food consumption, in: Public Health Nutrition 4(5B), 2001, p. 1159-1165

Naska A., Fouskakis D., Oikonomou E. (2005): Dietary patterns and their socio-demographic determinants in 10 European countries: data from the DAFNE databank, in: European Journal of Clinical Nutrition, November, 2005, p.1-10

REFERENCES

OECD (2001): Sustainable consumption: Sector case study series, ENV/EPOC/WPNEP, 13, 2001, p. 2-94

OECD (2006),http://stats.oecd.org/glossary/detail.asp (14/12/2006)

Paterakis S.E. and Nelson M. (2003): A comparison between the National Food Survey and the Family Expenditure Survey food expenditure data, in: Public Health Nutrition: 6(6), 2003, p.571-580

Payer H., Burger P., Lorek S. (2000): Food consumption in Austria – Driving Forces and Environmental Impacts – National case study for the OECD Programme on Sustainable Consumption, Federal Ministry of Agriculture, Forestry, Environment and Water Management, 2000, p. 2-55

Pomerleau J., Lock K., McKee M. (2003): Discrepancies between ecological and individual data on fruit and vegetable consumption in fifteen countries, in: British Journal of Nutrition, 89, 2003, p.827-834

Prättälä R.S., Groth M.V., Oltersdorf U.S. (2003): Use of butter and cheese in 10 European countries – A case of contrasting educational differences, in: European Journal of Public Health; 13, 2003, p.124-132

Ratinger T. and Šlaisova J. (2001): Market Potential for Food in Central and Eastern Europe - Analysis of Food Consumption in Central and Eastern Europe: Relevance and Empirical Methods, Volume 13, Wissenschaftsverlag Vauk Kiel KG, 2001, p.174-177

Rimmer D.J. (2001): An overview of food eaten outside the home in the United Kingdom National Food Survey and the new Expenditure and Food Survey, in: Public Health Nutrition 4(5B), 2001, p. 1173 - 1175

Rodrigues SSP. and de Almeida MDV. (2001): Portuguese household food availability in 1990 and 1995, in: Public Health Nutrition 4(5B), 2001, p. 1167-1171

Roos G., Johansson L., Kasmel A. (2001): Disparities in vegetable and fruit consumption: European cases from the north to the south, in: Public Health Nutrition, Volume 4: Issue 1, 2001, p. 35-43

REFERENCES

Rowely R. (2006): Policy Briefing: Implications of class based social and cultural aspects of food choice for local Health Promotion in Sandwell www.rrt-pct.org.uk/healthy-living/Implications-of-class-based (14/12/2006)

Sanchez-Villegas A., Martinez JA, Prättälä R. (2003): Systematic review of socioeconomic differences in food habits in Europe: consumption of cheese and milk, in: European Journal of Clinical Nutrition 57, 2003, p.917-929

Schulze A. (1998): Vergleich der Ernährungssituation in Deutschland und Großbritannien, Studien zur Haushaltsökonomie, Frankfurt am Main: Peter Lang-Verlag, Band 13, 1998, p.67-77

Sekula W., Lagiou P., Morawska M. (1997): A comparison of the food and health patterns in Greece and in Poland, in: Prace Oryginalne – Zywienie Czlowieka I Metabolizm, XXIV, nr1, 1997, p.3-15

Serra-Majem L. (2001): Food Availability and consumption at national, household and individual levels: implications for food-based dietary guidelines development, in: Public Health Nutrition, Volume 4: Issue 2(B), 2001, p.673-676

Serra-Majem L., MacLean D., Ribas L. (2003): Comparative and analysis of nutrition data from national, household, and individual levels: results from a WHO-CINDI collaborative project in Canada, Finland, Poland, and Spain, in: J Epidemiol Community Health, 57, 2003, p.74-80

Société française (2000) de santé publique: Health and Human Nutrition – Elements for European Action: French Presidency, 2,rue du Doyen Jacques Parisot – BP7, 54501 Vandoeuvre-lés-Nancy cedex, France, 2000, p.1-33

Stadt Wien (2002): Mikrozensus 1999 – Ergebnisse zur Gesundheit in Wien / Microcensus 1999 – Results on Health in Vienna, Studie S2/2002, bgf, 2002, p. 1-56

Statistik Austria (1999a): Konsumerhebung 1999 – Interviewerhandbuch, STATISTIK AUSTRIA – Die Informationsmanager, 1999, p. 3-63

REFERENCES

Statistik Austria (1999b): Konsumerhebung 1999 – Haushaltsbuch, STATISTIK AUSTRIA – Die Informationsmanager, 1999, p. 1-50

Statistik Austria (2001): Verbrauchsausgaben 1999/00 – Hauptergebnisse der Konsumerhebung, STATISTIK AUSTRIA – Die Informationsmanager, 2001, p. 1-336

Statistik Austria (2002): Verbrauchsausgaben 1999/00 – Sozialstatistische Ergebnisse der Konsumerhebung, STATISTIK AUSTRIA– Die Informationsmanager, 2002, p. 1-60

Statistik Austria (2003a): Qualitätsbericht zur Konsumerhebung 1999/2000 samt Methodenbeschreibung, STATISTIK AUSTRIA – Die Informationsmanager, 2003, p. 3-28

Statistik Austria (2003b): COICOP – HBS 2003, *(Classification of Individual Consumption Expenditures by Purpose)*, STATISTIK AUSTRIA – Die Informationsmanager, 2003

Statistik Austria (2004): Standard-Dokumentation Metainformationen – Definitionen, Erläuterungen, Methoden, Qualität zur Konsumerhebung 1999/2000, STATISTIK AUSTRIA – Die Informationsmanager, 2004, p. 2-29

Statistik Austria (2006): http://statistic.gv.at/English/results/population/families_txt.shtml (14/12/2006)

Szponar L., Sekula W., Nelson M. (2001): The Household Food Consumption and Anthropometric Survey in Poland, in: Public Health Nutrition 4(5B), 2001, p.1183-1186

Tippett K.S. (1999): Food consumption surveys in the US Department of agriculture, in: Nutrition Today, Jan, 1999, p. 1-12

Trichopoulou A. (1992): Monitoring food intake in Europe: A food databank based on Household Budget Surveys, in: European Journal of Clinical Nutrition, 46 – Suppl. 5, 1992, p. 3-8

REFERENCES

Trichopoulou A., Kanellou A., Lagiou P. (1996): Integration of nutritional data based on household budget surveys in European countries, in: Proceedings of the Nutrition Society, 55, 1996, p.699-704

Trichopoulou A. and Lagiou P. (1997a): DAFNE I – Methodology for the exploitation of HBS food data and results on food availability in 5 European countries, European Commission-COST 99, EUR 17909 EN, 1997, p. VII- XI + p.1-142

Trichopoulou A. and Lagiou P. (1997b) : Healthy Traditional Mediterranean Diet: An Expression of Culture, History, and Lifestyle, in: Nutrition Reviews, Vol. 55, No. 11, 1997, p. 383-389

Trichopoulou A., Lagiou P., Nelson M. (1999): Food disparities in 10 European countries: their detection using household budget survey data – The data Food Networking (DAFNE) Initiative (Statistical Data Included), in: Nutrition Today- May, 34:3, 1999, p.129-139

Trichopoulou A. and Lagiou P. (1999): DAFNE II – Methodology for the exploitation of HBS food data and results on food availability in six European countries, European Commission, EUR 18357 EN, 1999, p. 2-149

Trichopoulou A. and Naska A. (2001): Introduction, in: Public Health Nutrition 4(5B), 2001, p.1131-1132

Trichopoulou A. (2001): The DAFNE databank as a simple tool for nutrition policy, in: Public Health Nutrition 4(5B), 2001, p.1187-1198

Trichopoulou A. (2002): Data Food Networking – The European Food Availability Databank based on Household Budget Survey – The DAFNE IV Project – Technical Annex, Greece: University Athens, SPC.2002336, 2002, p. 2-17

Trichopoulou A., Naska A., Costacou T. (2002): Disparities in food habits across Europe, in: Proceedings of the Nutrition Society, Volume 61(4), 2002, p.553-558

REFERENCES

Trichopoulou A. and Naska A. (2003a): European food availability databank based on household budget surveys – The Data Food Networking initiative, in: European Journal of Public Health; 13 (3 Supplement), 2003, p.24-28

Trichopoulou A. and Naska A. (2003b): DAFNE III – Network for the Pan-European food data bank based on Household Budget Surveys, European Commission – Health Monitoring Programme of DG-SANCO, 2003, p. 3-42

Trichopoulou A., Naska A., Oikonomou E. (2005): The DAFNE databank: the past and future of monitoring the dietary habits of Europeans, in: Public Health, 13, 2005, p.69-73

Trichopoulou A. (2005): DAFNE IV – Network for the Pan-European food data bank based on Household Budget Surveys, European Commission – Health Monitoring Programme of DG-SANCO, 2005, p. 1-11

Turrell G., Blakely T., Patterson C. (2004): A multilevel analysis of socioeconomic (small area) differences in household food purchasing behaviour, in: Journal of Epidemiology and Community Health, 58, 2004, p.208-215

Unwin Ian (2000): Eurocode 2 Proposal Report P002 – The Euro Food Groups proposal – Cost Action 99 Report, Luxembourg: European Commission, 2000, p. 1-17

Vasdekis VGS., Stylianou S., Naska A. (2001): Estimation of age- and gender-specific food availability from household budget survey data, in: Public Health Nutrition 4(5B), 2001, p. 1149-1151

Zintzaras E., Kanellou A., Trichopoulou A. (1997): The validity of household budget survey (HBS) data: estimation of individual food availability in an epiemiological context, in: Journal of Human Nutrition and Dietetics, 10, 1997, p.53-62

Zunft H.-J., Ulbricht G., Pokorny J. (1999): Nutrition, physical activity and health status in Middle and East European countries, in: Public Health Nutrition, 2(3a), 1999, p.437-44

ANNEXES

ANNEX 1 Graphs of food and beverage availability by DAFNE food groups in Europe

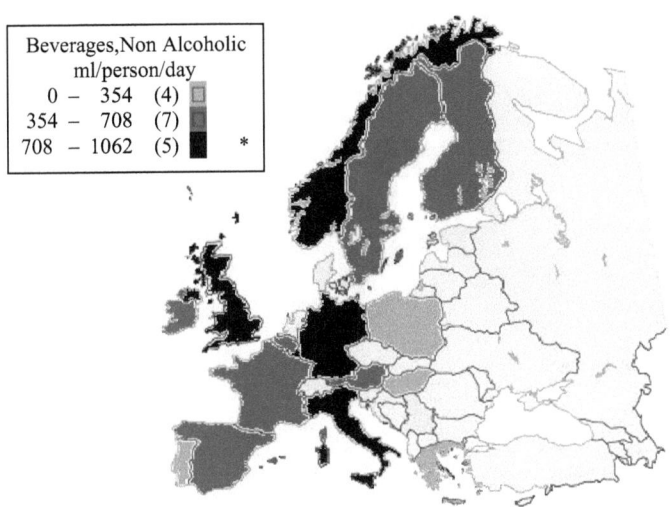

*source: dafne databank

ANNEX 1

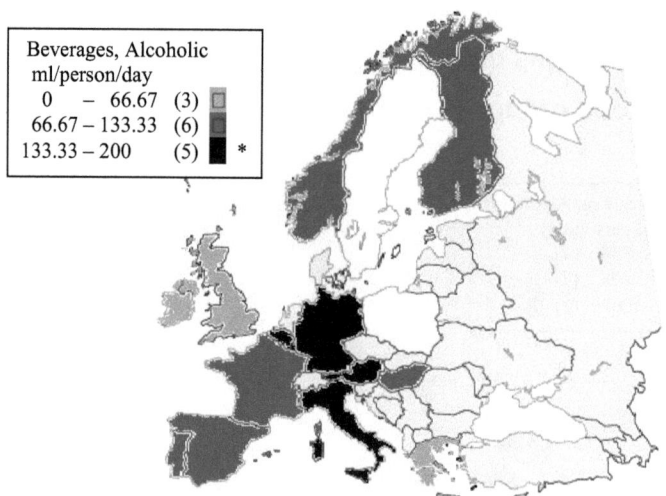

*source: dafne databank

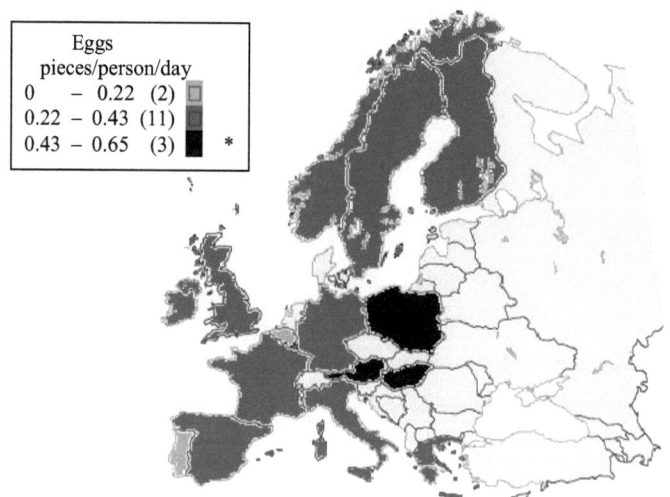

*source: dafne databank

ANNEX 1

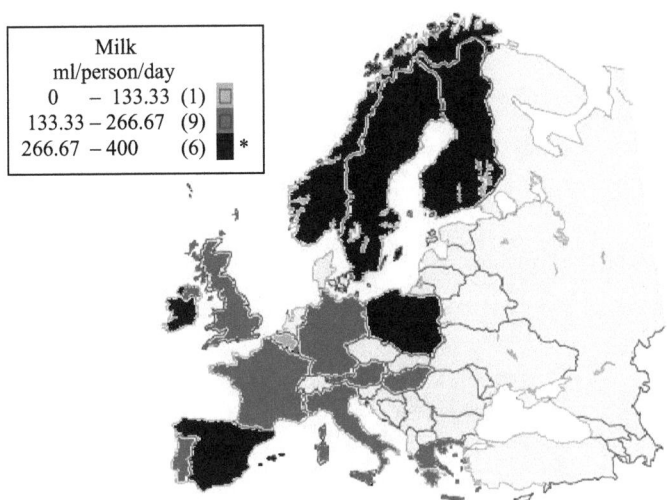

*source: dafne databank

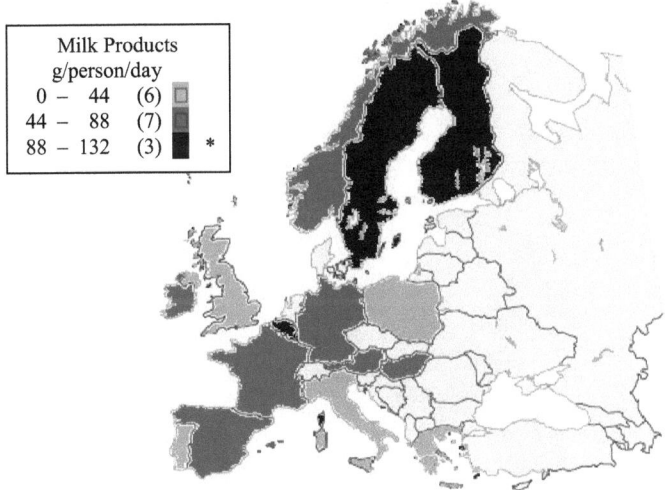

*source: dafne databank

ANNEX 1

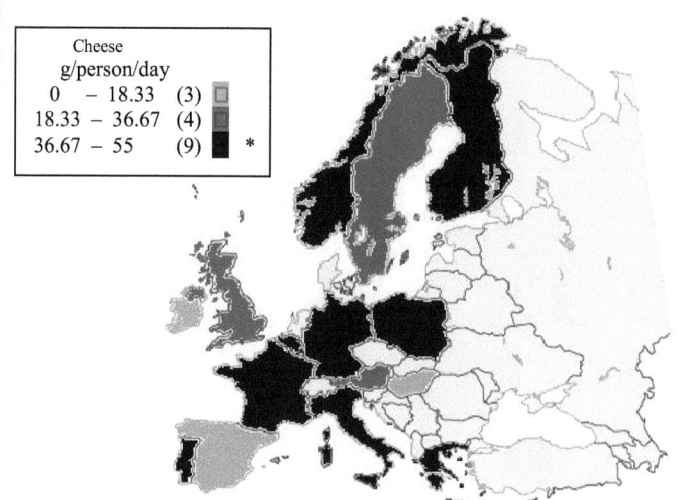

*source: dafne databank

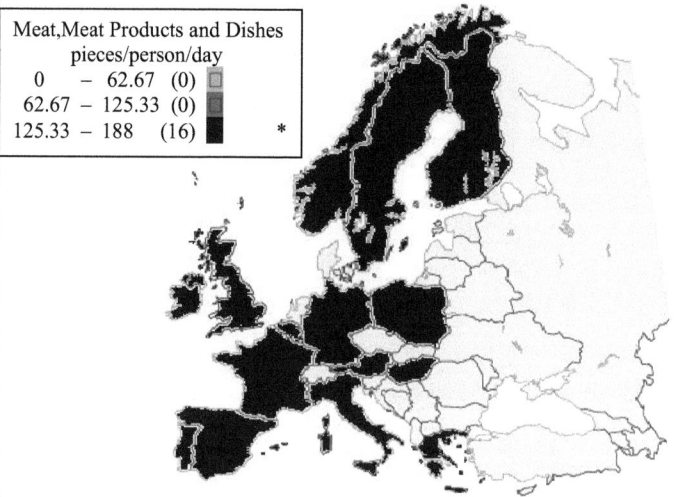

*source: dafne databank

ANNEX 1

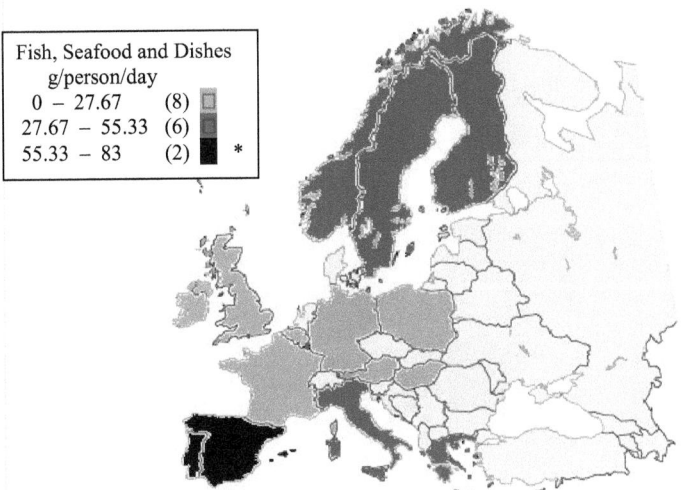

*source: dafne databank

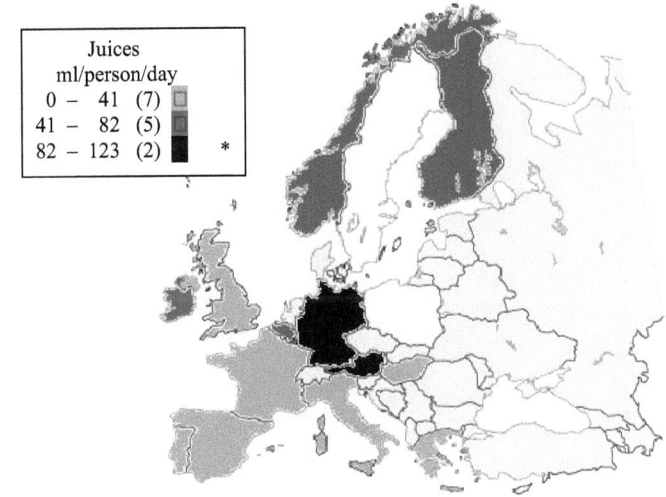

*source: dafne databank

ANNEX 1

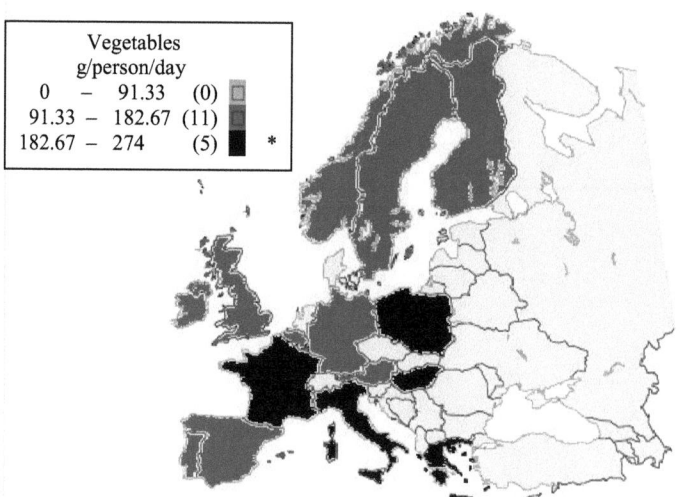

*source: dafne databank

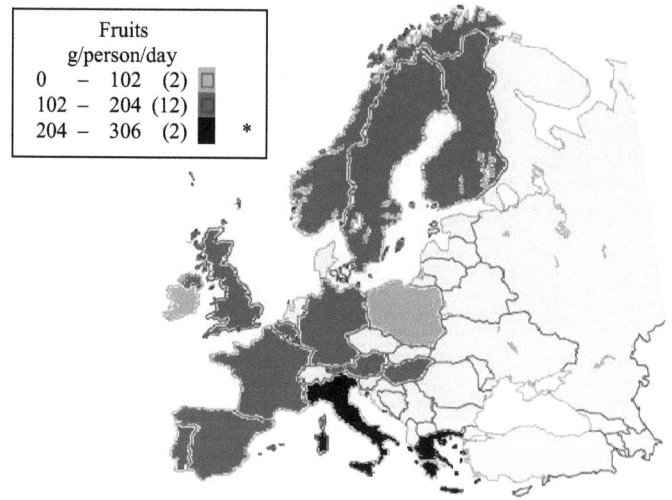

*source: dafne databank

ANNEX 1

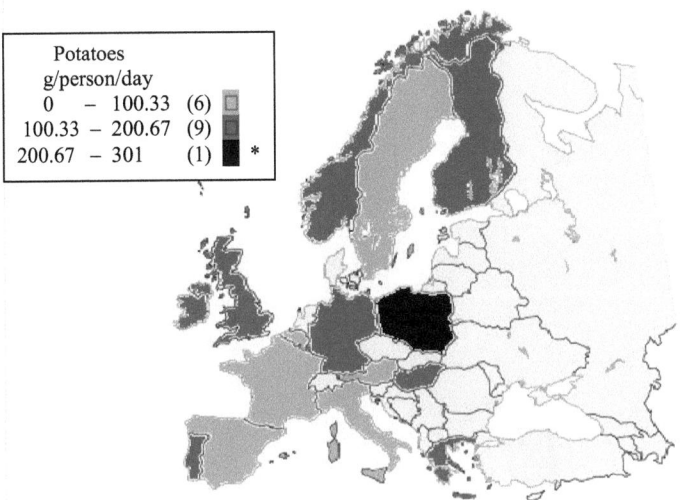

*source: dafne databank

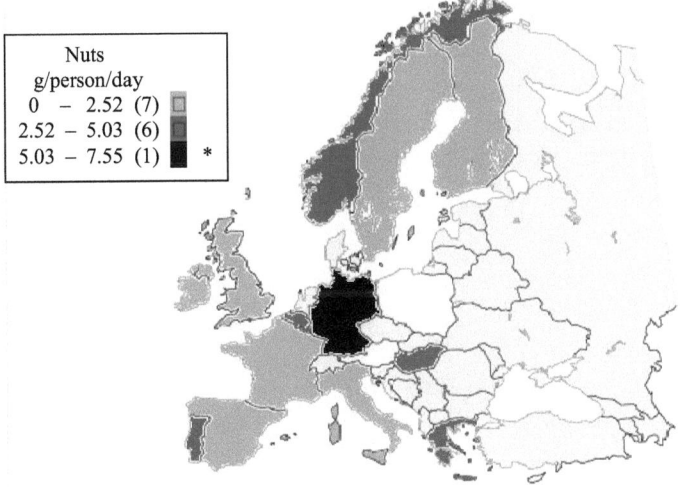

*source: dafne databank

ANNEX 1

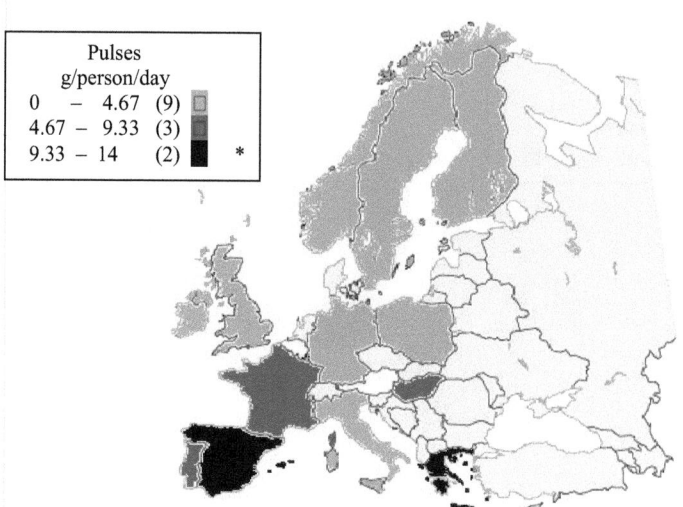

*source: dafne databank

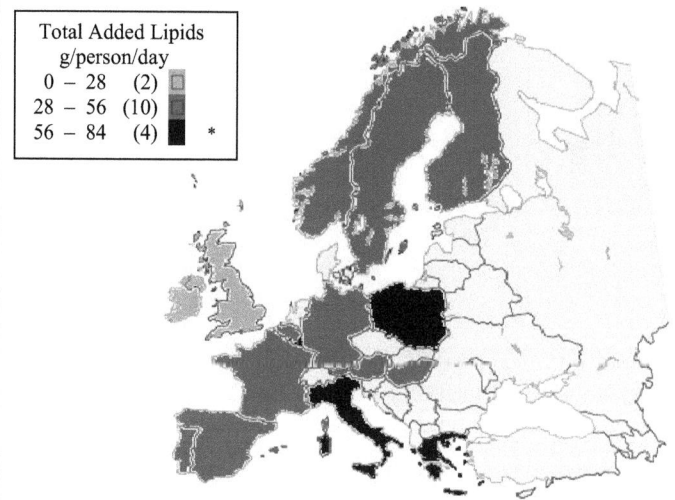

*source: dafne databank

ANNEX 1

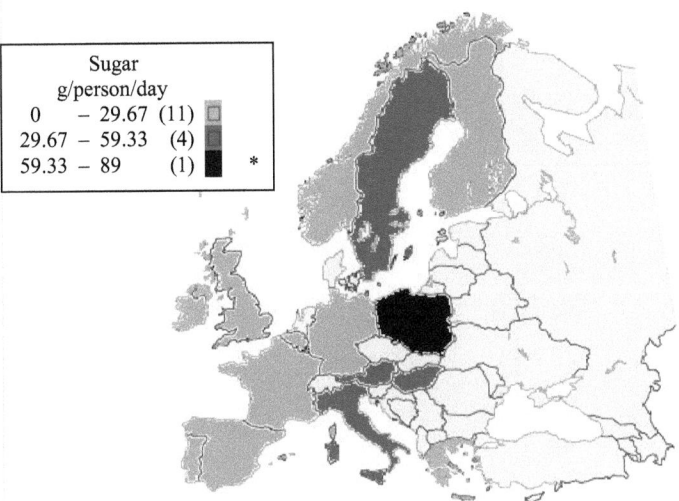

*source: dafne databank

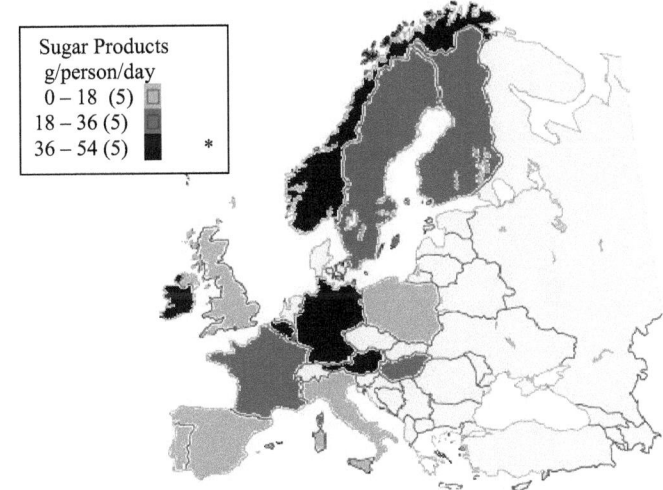

*source: dafne databank

ANNEX 1

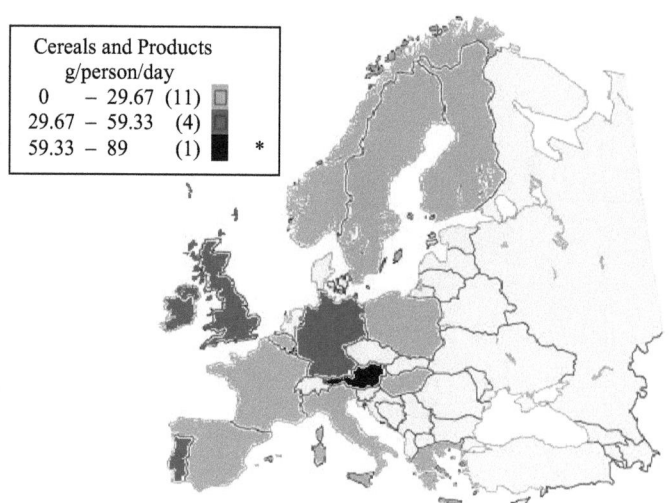

*source: dafne databank

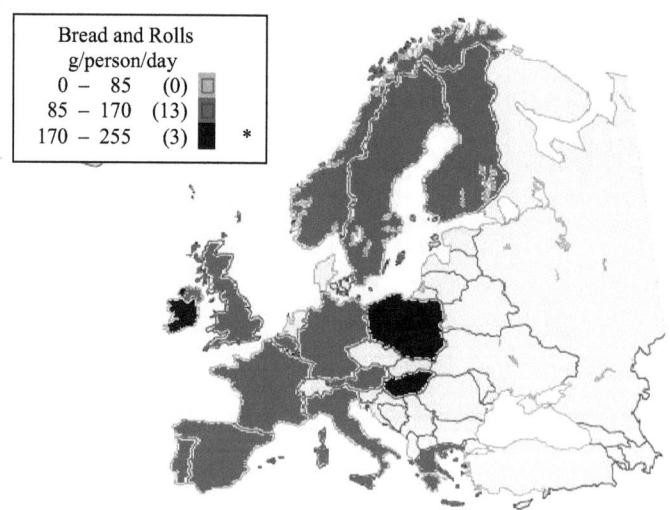

*source: dafne databank

ANNEX 1

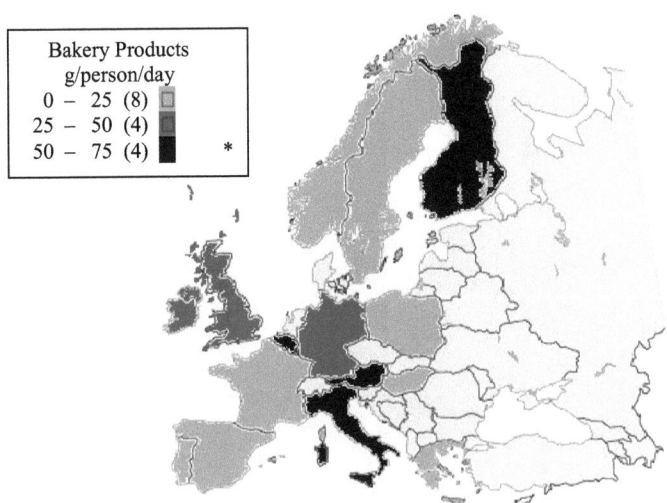

*source: dafne databank

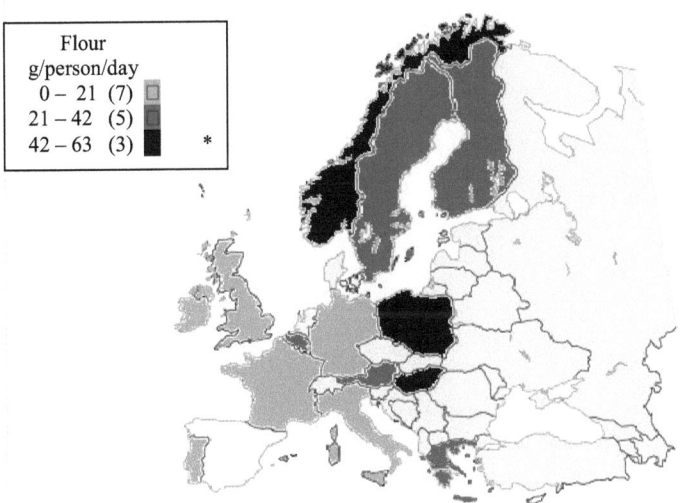

*source: dafne databank

A11

ANNEX 1

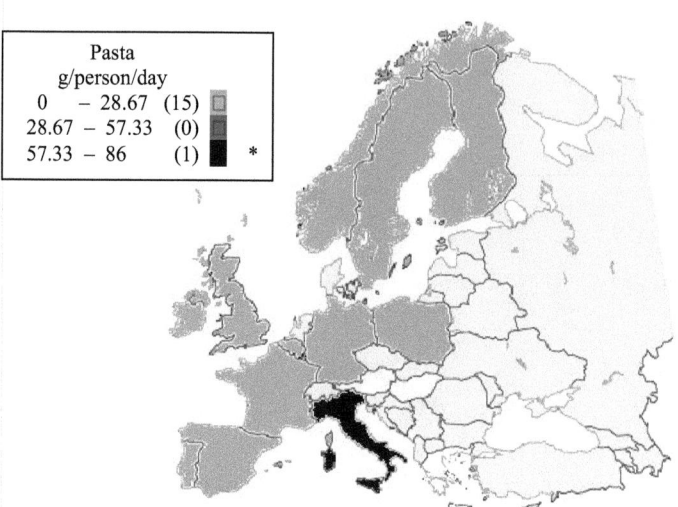

*source: dafne databank

ANNEX 2: DETAILED TABLES OF RESULTS

ANNEX 2.1: Tables of food availability by locality
AUSTRIA

BEVERAGES AND STIMULANTS person/day

food group	rural	semi-urban	urban
Mineral Water (ml)	214	207	206
Lemonades (ml)	106	124	122
Coffee (g)	15	17	18
Cocoa (g)	3,11	2,53	2,55
Beer (ml)	123	146	93
White wine (ml)	17	18	20
Red wine, Rose (ml)	11	16	18
Other fruit wine (ml)	0,82	1,13	0,94
Sparkling wine, champagne, vermouth (ml)	17	15	10
Schnapps, spirits, liqueurs (ml)	4,08	4,63	4,33

FISH AND SEAFOOD g/person/day

food group	rural	semi-urban	urban
Fish (fresh or frozen)	6,80	7,75	8,14
Fish, seafood (dried, smoked)	0,36	0,70	1,06
Preserved / processed fish or seafood	0,54	0,72	0,77
Seafood (fresh or frozen)	0,18	0,47	0,76

FATS AND OILS g/person/day

food group	rural	semi-urban	urban
Vegetable fats (g)	12	11	9,74
Olive Oil (ml)	2,41	2,98	4,42
Vegetable Oils (ml) (excluding olive oils)	19	20	16
Butter	12	11	9,95
Lipids of animal origin (butter excluded)	1,76	1,23	0,70

JUICES ml/person/day

food group	rural	semi-urban	urban
Fruit juices	72	84	109
Vegetable juices	1,15	1,81	3,36

EGGS, MILK AND MILK PRODUCTS g/person/day

food group	rural	semi-urban	urban
Eggs (pieces)	0,56	0,50	0,42
Cheese	17	18	21
Curd cheese	9,38	9,50	9,69
Fresh milk	186	147	145
Preserved milk	28	34	33
Yoghurt	32	36	46
Other milk products	12	14	15
Ice cream (ml)	16	16	22

FRUITS AND NUTS g/person/day

food group	rural	semi-urban	urban
Apples	73	70	49
Bananas	27	30	32
Berry fruits (excl. grapes)	12	15	13
Cherries	3,67	3,55	2,98
Citrus fruits	28	34	36
Dried fruits	4,36	4,78	5,85
Grapes	5,33	6,76	5,83
Other fruits	9,35	13	19
Peaches,nectarines,apricots	12	11	9,46
Pears	7,46	7,75	7,98
Plums	6,27	6,06	5,08
Preserved,frozen fruits	0,96	0,96	1,87

ANNEX 2

CEREALS AND CEREAL PRODUCTS   g/person/day

food group	rural	semi-urban	urban
Dark bread	72	67	59
White bread	22	23	29
Biscuits	27	29	28
Cakes and pastries	11	15	21
Cookies,biscuits,wafers	15	17	21
Rusks,crispbread	0,99	1,38	1,35
Flour	54	41	23
Pasta	19	20	23
Rice	12	14	18
Other cereal products	62	74	91

MEAT AND MEAT PRODUCTS   g/person/day

food group	rural	semi-urban	urban
Poultry (fresh or frozen)	23	22	23
Pork (fresh or frozen)	45	31	21
Minced meat (50% pork)	4,57	5,62	5,36
Beef meat (fresh or frozen)	19	15	14
Veal meat (fresh or frozen)	1,49	2,15	2,24
Minced meat (50% beef)	4,57	5,62	5,36
Sheep&goat meat (fresh or frozen)	1,85	0,54	2,35
Other meat (fresh or frozen)	1,86	1,71	1,45
Sausages	37	36	32
Other preserved or processed meat products	37	52	49
Smoked products	12	10	8,16
Spread meat	1,24	1,53	1,54
Offals	1,49	1,59	2,27

SUGAR AND SUGAR PRODUCTS   g/person/day

food group	rural	semi-urban	urban
Sugar, sweetener	46	33	23
Sweets	6,52	6,96	8,52
Chocolate	18	21	29
Jam.honey	9,11	8,84	9,95

ANNEX 2

POTATOES   g/person/day

food group	rural	semi-urban	urban
Potatoes	95	85	65
Tuber plants and products	15	19	22

VEGETABLES   g/person/day

food group	rural	semi-urban	urban
Tomato	19	21	22
Stem vegetables (excl. tomatoes)	19	21	23
Cabbage	15	16	16
Carrots	8,93	9,41	9,51
Leafy,vegetables,herbs	32	31	28
Onions	16	17	17
Root vegetables,mushrooms (excl. onions&carrots)	1,05	1,11	1,12
Dried vegetables	0,98	1,62	1,34
Preserved,frozen vegetables	22	29	32

ANNEX 2.2: Tables of food availability by education
AUSTRIA

BEVERAGES AND STIMULANTS   person/day

food group	elementary	secondary	higher
Mineral Water (ml)	208	213	172
Lemonades (ml)	121	119	86
Coffee (g)	20	17	16
Cocoa (g)	1,26	2,88	1,83
Beer (ml)	130	122	86
White wine (ml)	8,53	18	19
Red wine, Rose (ml)	6,61	14	20
Other fruit wine (ml)	0,15	0,96	0,33
Sparkling wine, champagne, vermouth (ml)	23	14	12
Schnapps, spirits, liqueurs (ml)	4,50	4,35	3,89

CEREALS AND CEREAL PRODUCTS   g/person/day

food group	elementary	secondary	higher
Dark bread	77	67	58
White bread	23	25	22
Biscuits	24	27	29
Cakes and pastries	14	15	18
Cookies, biscuits, wafers	15	18	20
Rusks, crispbread	1,68	1,15	1,84
Flour	98	42	15
Pasta	16	20	22
Rice	16	14	15
Other cereal products	54	72	106

ANNEX 2

EGGS, MILK AND MILK PRODUCTS   g/person/day

food group	elementary	secondary	higher
Eggs (pieces)	0,59	0,51	0,31
Cheese	17	18	21
Curd cheese	9,12	9,57	9,17
Fresh milk	205	164	134
Preserved milk	46	32	21
Yoghurt	39	37	51
Other milk products	16	14	15
Ice cream (ml)	18	17	25

FATS AND OILS   g/person/day

food group	elementary	secondary	higher
Vegetable fats	12	11	6,22
Olive oil	2,40	3,13	4,15
Other salad, cooking oil	42	19	10
Butter	13	11	11
Animal fat	1,08	1,35	0,42

FISH AND SEAFOOD   g/person/day

food group	elementary	secondary	higher
Fish (fresh or frozen)	6,35	7,27	9,93
Fish, seafood (dried, smoked)	0,00	0,68	0,73
Preserved or processed fish or seafood	0,63	0,66	0,73
Seafood (fresh or frozen)	0,09	0,42	0,78

ANNEX 2

FRUITS   g/person/day

food group	elementary	secondary	higher
Apples	116	64	57
Bananas	36	29	29
Berry fruits (excl. grapes)	15	13	12
Cherries	5,20	3,40	3,08
Citrus fruits	33	32	34
Grapes	6,52	5,89	5,35
Peaches,nectarines,apricots	17	11	9,77
Other fruits	11	13	17
Pears	8,31	7,61	8,44
Plums	8,87	5,80	5,25
Dried fruits	4,47	4,72	7,53
Preserved,frozen fruits	0,45	1,18	2,41

VEGETABLES   g/person/day

food group	elementary	secondary	higher
Tomato	20	20	25
Stem vegetables (excl. tomatoes)	20	21	26
Cabbage	42	15	14
Carrots	9,66	9,04	11
Leafy,vegetables,herbs	52	30	26
Onions	18	16	20
Root vegetables,mushrooms (excl. onions&carrots)	1,14	1,06	1,28
Dried vegetables	4,85	1,23	1,18
Preserved,frozen vegetables	26	26	35

ANNEX 2

MEAT AND MEAT PRODUCTS   g/person/day

food group	elementary	secondary	higher
Poultry (fresh or frozen)	28	23	18
Pork (fresh or frozen)	32	35	15
Minced meat (50% pork)	3,03	5,29	3,48
Beef meat (fresh or frozen)	31	16	12
Veal meat (fresh or frozen)	3,01	1,90	1,55
Minced meat (50% beef)	3,03	5,29	3,48
Sheep&goat meat (fresh or frozen)	7,42	1,65	1,22
Other meat (fresh or frozen)	0,99	1,72	1,38
Sausages	33	36	27
Other preserved or processed meat products	39	45	51
Smoked products	13	11	4,66
Spread meat	1,37	1,45	1,13
Offals	1,90	1,85	0,88
Poultry (fresh or frozen)			

POTATOES   g/person/day

food group	elementary	secondary	higher
Potatoes	134	85	49
Tuber plants and products	16	18	19

JUICES   ml/person/day

food group	elementary	secondary	higher
Fruit juices	94	85	114
Vegetable juices	0,25	1,89	3,55

SUGAR AND SUGAR PRODUCTS   g/person/day

food group	elementary	secondary	higher
Sugar, sweetener	50	36	15
Sweets	7,11	7,27	7,73
Chocolate	22	22	24
Jam, honey	18	9,23	9,22

ANNEX 2.3: Tables of food availability by occupation
AUSTRIA

BEVERAGES AND STIMULANTS   g/person/day

food group	manual	non manual	retired	unemployed	others
Mineral Water (ml)	197	197	257	164	162
Lemonades (ml)	134	118	92	128	138
Coffee (g)	16	15	22	17	11
Cocoa (g)	3,61	2,43	2,31	4,35	3,68
Beer (ml)	134	92	154	154	97
White wine (ml)	13	18	27	7,61	9,88
Red wine, Rose (ml)	8,42	15	20	11	7,09
Other fruit wine (ml)	1,00	0,80	1,17	0,54	0,88
Sparkling wine, champagne, vermouth (ml)	12	13	21	15	4,29
Schnapps, spirits, liqueurs (ml)	2,68	3,37	7,63	3,65	3,54

CEREALS AND CEREAL PRODUCTS   g/person/day

food group	manual	non manual	retired	unemployed	others
Dark bread	62	55	92	60	52
White bread	25	23	27	33	21
Biscuits	26	29	28	22	25
Cakes and pastries	12	15	19	14	13
Cookies, biscuits, wafers	15	18	20	14	20
Rusks, crispbread	0,97	1,11	1,69	1,29	0,86
Flour	42	29	60	24	29
Pasta	18	20	24	21	19
Rice	15	12	18	17	11
Other cereal products	67	80	69	106	76

EGGS, MILK AND MILK PRODUCTS   g/person/day

food group	manual	non manual	retired	unemployed	others
Eggs (pieces)	0,49	0,39	0,70	0,45	0,39
Cheese	16	20	20	19	13
Curd cheese	7,91	9,05	13	9,18	6,10
Fresh milk	142	147	207	148	167
Preserved milk	34	24	40	49	33
Yoghurt	33	39	41	33	38
Other milk products	11	14	16	12	12
Ice cream (ml)	18	19	16	12	18

FATS AND OILS   g/person/day

food group	manual	non manual	retired	unemployed	others
Vegetable fats	10	8,47	16	9,45	7,95
Olive oil	2,40	3,10	4,47	5,53	1,82
Other salad, cooking oil	20	13	26	22	20
Butter	9,02	9,29	16	8,53	7,46
Animal fat	1,32	0,63	2,42	1,73	0,66

FISH AND SEAFOOD   g/person/day

food group	manual	non manual	retired	unemployed	others
Fish (fresh or frozen)	5,61	6,87	10	8,48	6,76
Fish, seafood (dried, smoked)	0,33	0,91	0,69	0,26	0,52
Preserved or processed fish or seafood	0,58	0,67	0,73	0,90	0,57
Seafood (fresh or frozen)	0,33	0,52	0,39	0,54	0,73

FRUITS   g/person/day

food group	manual	Non manual	retired	unemployed	others
Apples	45	48	111	56	55
Bananas	29	29	30	29	32
Berry fruits (excl. grapes)	11	11	20	12	8,95
Cherries	2,62	2,99	5,16	3,41	1,95
Citrus fruits	28	30	43	32	21
Grapes	4,88	4,94	8,79	5,38	4,02
Peaches,nectarines,apricots	8,33	9,50	16	11	6,18
Other fruits	12	14	15	14	12
Pears	6,37	6,04	12	6,72	5,17
Plums	4,47	5,10	8,80	5,81	3,32
Dried fruits	3,26	4,88	7,08	3,63	3,79
Preserved,frozen fruits	0,90	1,47	1,14	1,35	1,61

SUGAR AND SUGAR PRODUCTS   g/person/day

food group	manual	non manual	retired	unemployed	others
Sugar, sweetener	30	24	59	34	24
Sweets	7,73	7,28	6,85	8,20	7,50
Chocolate	20	23	24	15	23
Jam.honey	7,42	6,90	16	6,68	7,51

ANNEX 2

MEAT AND MEAT PRODUCTS   g/person/day

food group	manual	non manual	retired	unemployed	others
Poultry (fresh or frozen)	22	20	30	26	20
Pork (fresh or frozen)	31	24	52	39	30
Minced meat (50% pork)	5,38	4,36	6,19	6,74	4,48
Beef meat (fresh or frozen)	16	13	24	17	9,49
Veal meat (fresh or frozen)	2,05	1,61	2,54	5,01	0,41
Minced meat (50% beef)	5,38	4,36	6,19	6,74	4,48
Sheep&goat meat (fresh or frozen)	2,67	1,06	1,68	2,43	2,29
Other meat (fresh or frozen)	1,20	1,15	3,24	2,43	0,68
Sausages	35	32	41	34	33
Other preserved or processed meat products	35	41	63	48	33
Smoked products	8,44	7,71	18	7,74	5,27
Spread meat	1,39	1,38	1,57	1,71	1,06
Offals	1,42	1,26	3,24	2,68	0,56

POTATOES   g/person/day

food group	manual	non manual	retired	unemployed	others
Potatoes	72	57	139	98	62
Tuber plants and products	18	18	17	21	27

ANNEX 2

VEGETABLES
g/person/day

food group	manual	non manual	retired	unemployed	others
Tomato	17	19	27	22	14
Stem vegetables (excl. tomatoes)	18	20	28	23	15
Cabbage	12	12	25	12	10
Carrots	7,27	8,57	13	12	6,71
Leafy, vegetables, herbs					
Onions	13	16	23	22	12
Root vegetables, mushrooms (excl. onions&carrots)	0,85	1,01	1,49	1,44	0,79
Dried vegetables	0,96	1,03	1,79	3,66	1,49
Preserved, frozen vegetables	20	28	32	23	24

JUICES
ml/person/day

food group	manual	non manual	retired	unemployed	others
Fruit juices	86	92	78	121	95
Vegetable juices	1,02	2,13	2,65	2,48	1,94

ANNEX 2

ANNEX 2.4: Tables of food availability by household composition   AUSTRIA

BEVERAGES AND STIMULANTS   g/person/day

food group	adult single	adult 2 members	elderly single	elderly 2 members
Mineral Water (ml)	267	255	286	241
Lemonades (ml)	180	127	72	68
Coffee (g)	29	20	25	20
Cocoa (g)	1,20	2,12	2,92	1,69
Beer (ml)	161	151	108	157
White wine (ml)	32	23	22	32
Red wine,Rose (ml)	29	20	22	24
Other fruit wine (ml)	1,80	1,00	1,09	2,03
Sparkling wine,champagne,vermouth (ml)	14	17	20	9,11
Schnapps,spirits,liqueurs (ml)	5,57	6,10	11	7,19

CEREALS AND CEREAL PRODUCTS   g/person/day

food group	adult single	adult 2 members	elderly single	elderly 2 members
Dark bread	81	68	101	95
White bread	32	30	32	27
Biscuits	31	31	30	30
Cakes and pastries	23	18	24	19
Cookies,biscuits,wafers	20	22	25	20
Rusks,crispbread	1,66	1,53	2,44	2,05
Flour	28	37	58	65
Pasta	34	19	26	25
Rice	29	14	21	17
Other cereal products	91	68	65	58

EGGS, MILK AND MILK PRODUCTS   g/person/day

food group	adult single	adult 2 members	elderly single	elderly 2 members
Eggs (pieces)	0,57	0,52	0,80	0,60
Cheese	27	23	22	19
Curd cheese	10	11	15	13
Fresh milk	162	132	232	198
Preserved milk	51	40	46	35
Yoghurt	61	45	49	36
Other milk products	19	15	20	15
Ice cream (ml)	20	19	15	9,96

FATS AND OILS   g/person/day

food group	adult single	adult 2 members	elderly single	elderly 2 members
Vegetable fats	9,03	12	20	19
Olive oil	6,75	4,28	5,76	5,93
Other salad, cooking oil	19	19	32	23
Butter	11	12	21	15
Animal fat	0,62	1,46	2,35	2,87

FISH AND SEAFOOD   g/person/day

food group	adult single	adult 2 members	elderly single	elderly 2 members
Fish (fresh or frozen)	5,70	9,60	11	12
Fish, seafood (dried, smoked)	0,74	1,82	0,51	1,44
Preserved or processed fish or seafood	1,37	0,89	0,76	0,90
Seafood (fresh or frozen)	0,91	0,69	0,23	0,35

ANNEX 2

FRUITS   g/person/day

food group	adult single	adult 2 members	elderly single	elderly 2 members
Apples	88	84	117	90
Bananas	42	31	36	28
Berry fruits (excl. grapes)	15	16	25	17
Cherries	3,48	4,40	5,56	4,48
Citrus fruits	44	41	58	51
Grapes	6,71	7,04	11	7,73
Peaches,nectarines,apricots	11	14	18	14
Other fruits	17	20	18	17
Pears	8,52	8,70	21	12
Plums	5,93	7,50	9,49	7,64
Dried fruits	8,19	6,27	8,08	9,75
Preserved,frozen fruits	1,64	2,16	2,13	0,60

VEGETABLES   g/person/day

food group	adult single	adult 2 members	elderly single	elderly 2 members
Tomato	23	26	25	28
Stem vegetables (excl. tomatoes)	24	27	26	29
Cabbage	13	21	31	25
Carrots	11	11	12	13
Leafy,vegetables,herbs	28	34	59	44
Onions	20	20	23	23
Root vegetables,mushrooms (excl. onions&carrots)	1,28	1,31	1,46	1,50
Dried vegetables	1,45	1,25	1,93	2,06
Preserved,frozen vegetables	39	35	38	31

JUICES   ml/person/day

food group	adult single	adult 2 members	elderly single	elderly 2 members
Fruit juices	121	93	85	68
Vegetables juices	5,77	2,78	5,17	1,30

ANNEX 2

MEAT AND MEAT PRODUCTS

g/person/day

food group	adult single	adult 2 members	elderly single	elderly 2 members
Poultry (fresh or frozen)	25	26	28	33
Pork (fresh or frozen)	24	36	36	36
Minced meat (50% pork)	4,92	5,35	6,25	5,60
Beef meat (fresh or frozen)	13	19	23	20
Veal meat (fresh or frozen)	2,83	2,29	2,35	2,24
Minced meat (50% beef)	4,92	5,35	6,25	5,60
Sheep&goat meat (fresh or frozen)	2,10	2,13	0,93	0,56
Other meat (fresh or frozen)	2,29	2,09	2,52	1,89
Sausages	37	40	38	41
Other preserved or processed meat products	71	62	52	47
Smoked products	12	13	19	16
Spread meat	2,11	1,50	2,34	1,34
Offals	1,89	2,16	3,74	2,39

POTATOES

g/person/day

food group	adult single	adult 2 members	elderly single	elderly 2 members
Potatoes	67	83	145	129
Tuber plants and products	19	19	22	15

SUGAR AND SUGAR PRODUCTS

g/person/day

food group	adult single	adult 2 members	elderly single	elderly 2 members
Sugar, sweetener	27	33	74	48
Sweets	8,43	8,31	9,89	6,27
Chocolate	36	26	32	22
Jam.honey	14	12	17	17

ANNEX 3
ANNEX 3: Graphs on food availability by socio-demographic characteristics

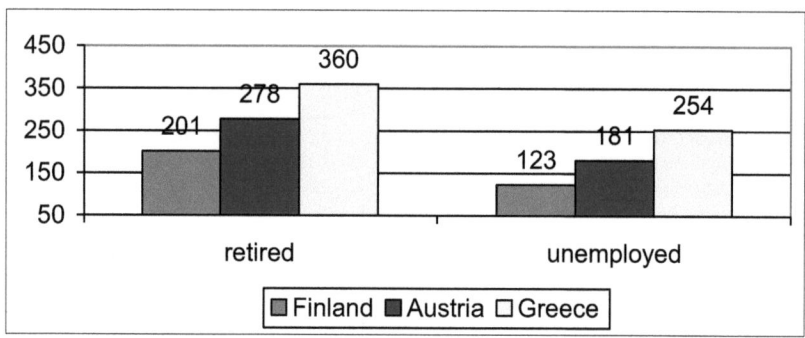

Figure 78. Daily availability of fruits, by occupation (g/p/d)

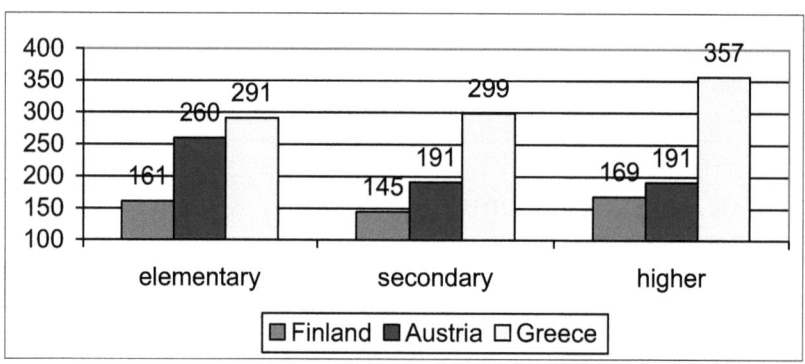

Figure 79. Daily availability of fruits, by education (g/p/d)

ANNEX 3

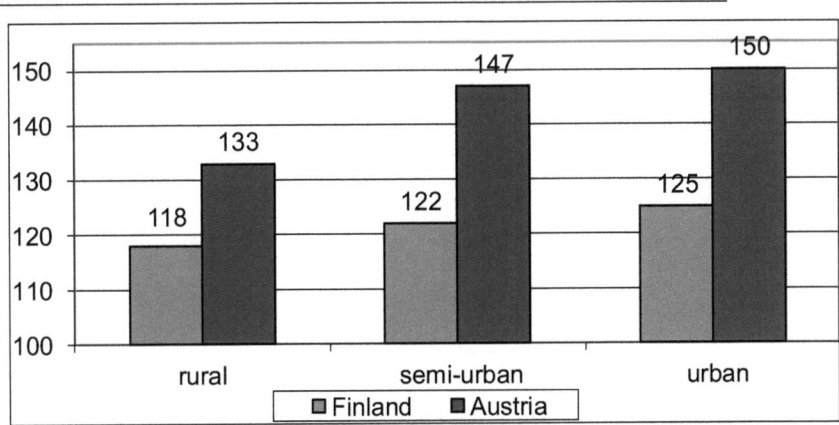

Figure 80. Daily availability of vegetables, by locality (g/p/d)

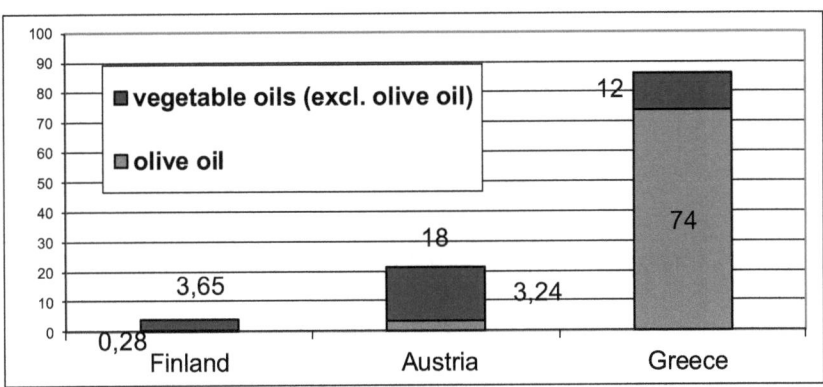

Figure 81. Daily availability of vegetable oils, by mean (ml/p/d)

ANNEX 3

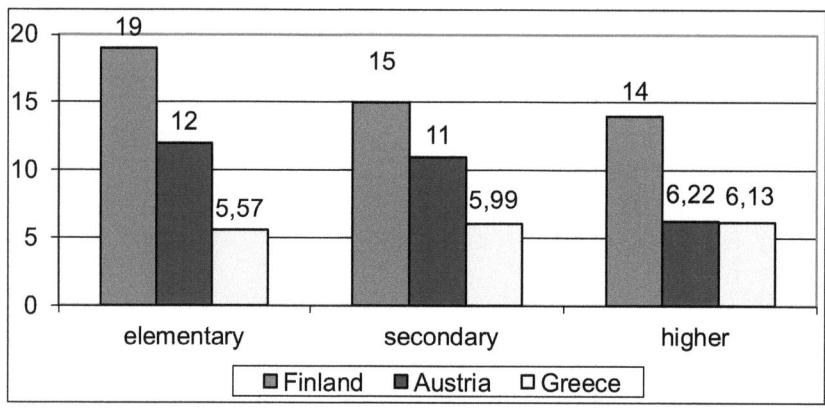

Figure 82. Daily Availability of vegetable fats, by education (g/p/d)

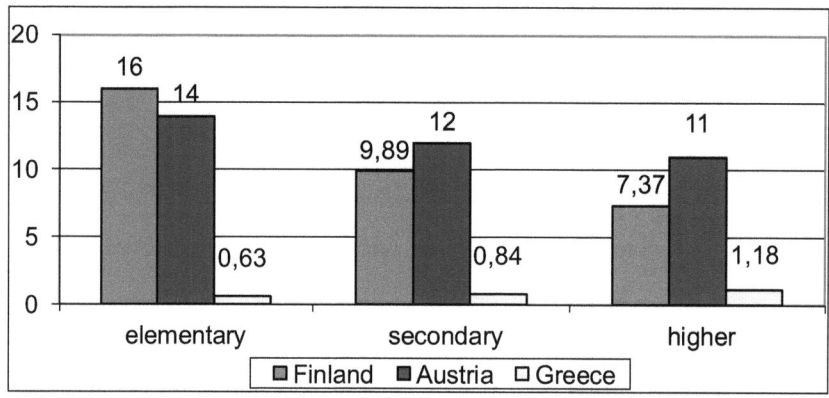

Figure 83. Daily availability of lipids of animal origin, by education (g/p/d)

A32

ANNEX 3

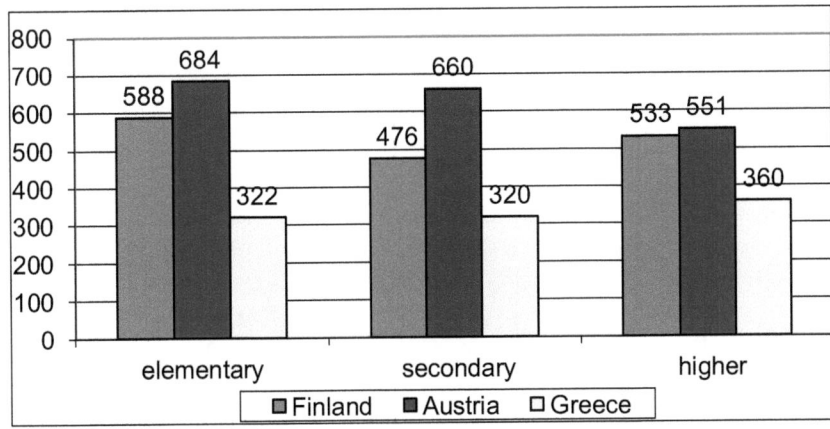

Figure 84. Daily availability of non alcoholic beverages,
by education (ml/p/d)

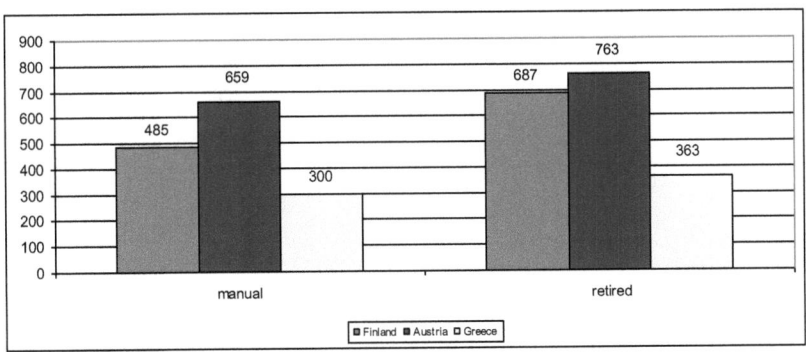

Figure 85. Daily availability of non alcoholic beverages,
by occupation (ml/p/d)

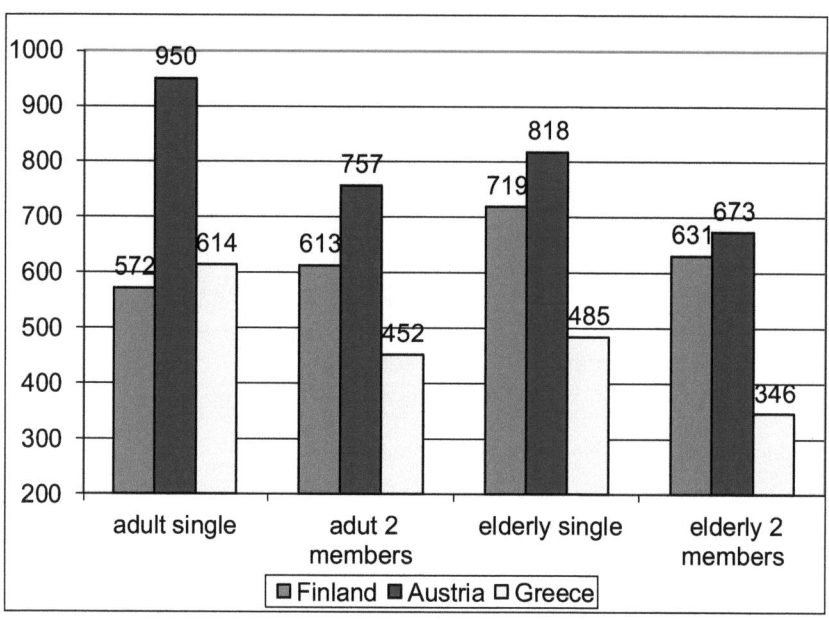

Figure 86. Daily availability of non alcoholic beverages, by household composition (ml/p/d)

ANNEX 3

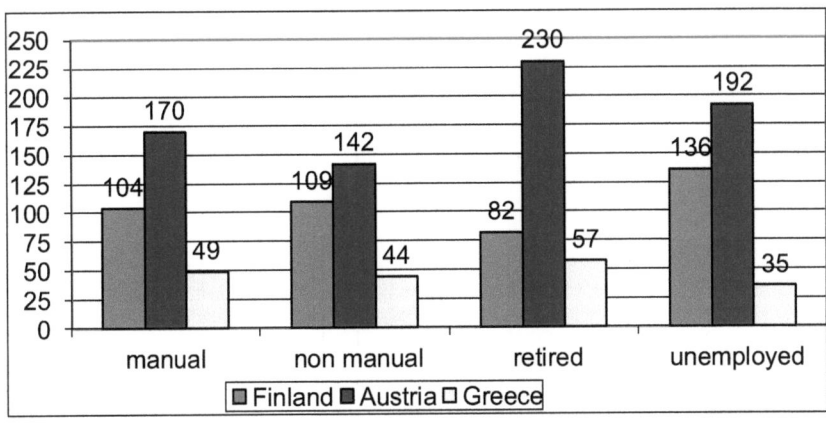

Figure 87. Daily availability of alcoholic beverages by occupation (ml/p/d)

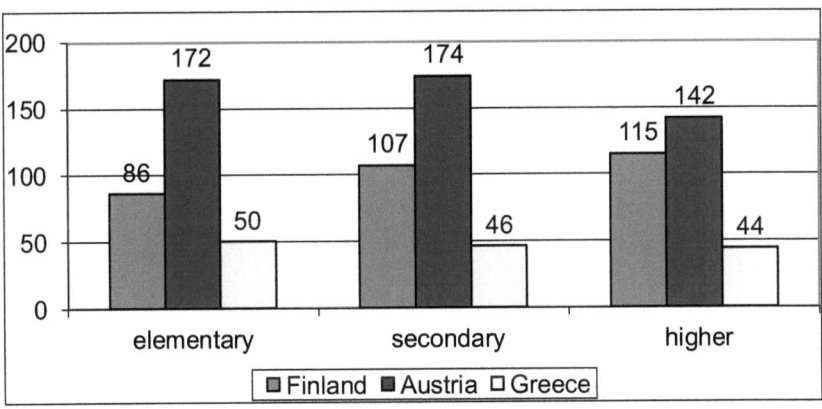

Figure 88. Daily availability of alcoholic beverages, by education (ml/p/d)

A35

ANNEX 3

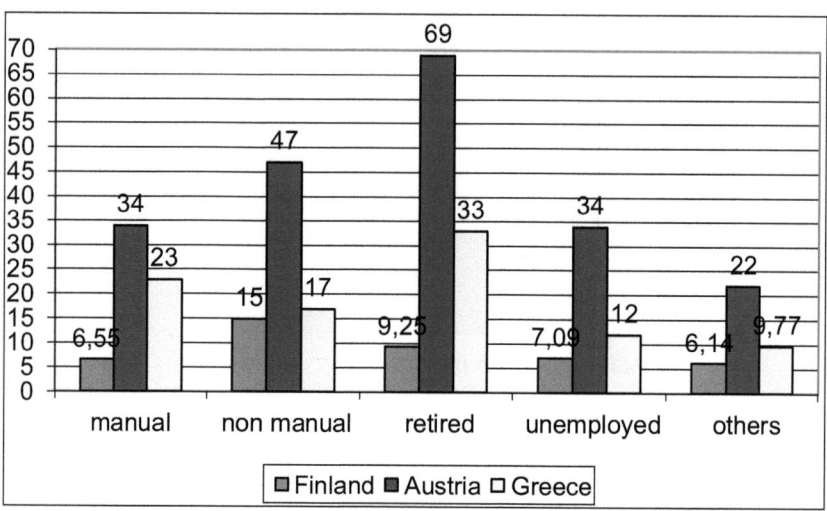

Figure 89. Daily availability of wine, by occupation (ml/p/d)

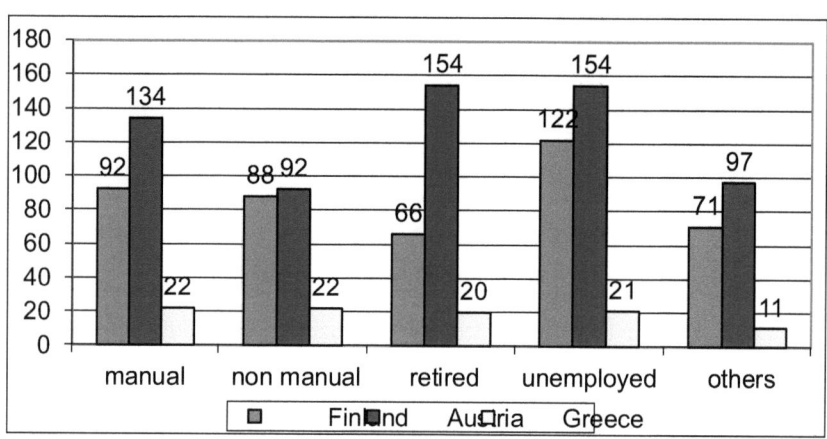

Figure 90. Daily availability of beer, by occupation (ml/p/d)

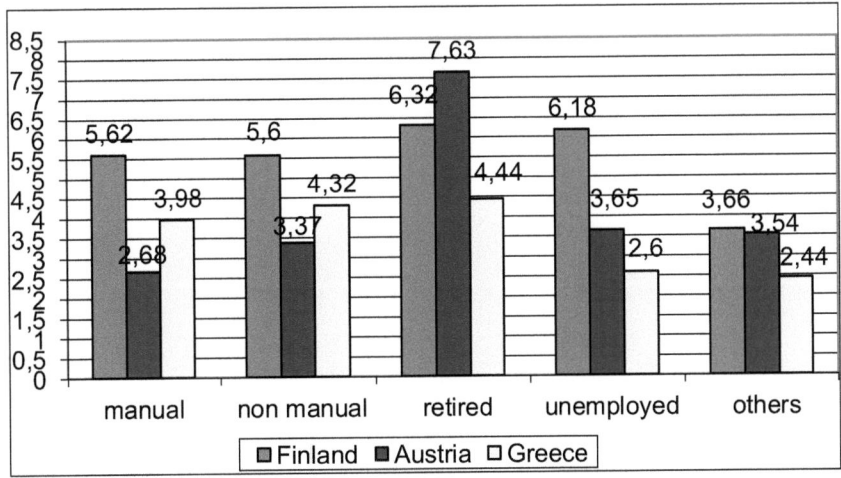

Figure 91. Daily availability of spirits, by occupation (ml/p/d)

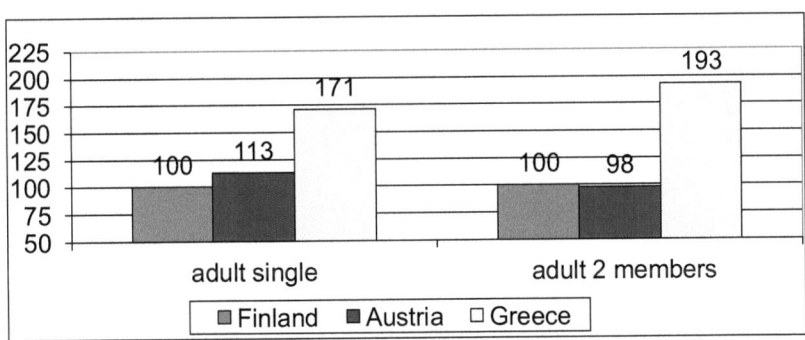

Figure 92. Daily availability of bread and rolls, by household composition (g/p/d)

ANNEX 3

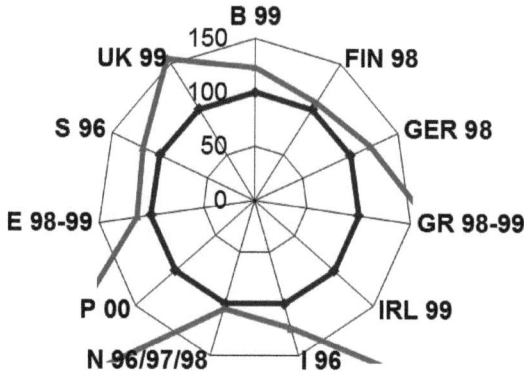

Figure 93a. Potatoes, by "adult-single"

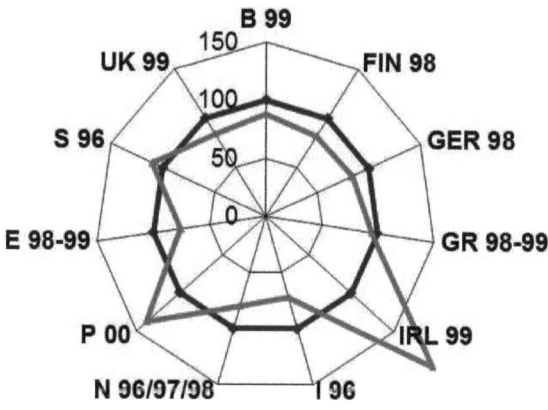

Figure 93b. Potatoes, by "elderly-single

ANNEX 3

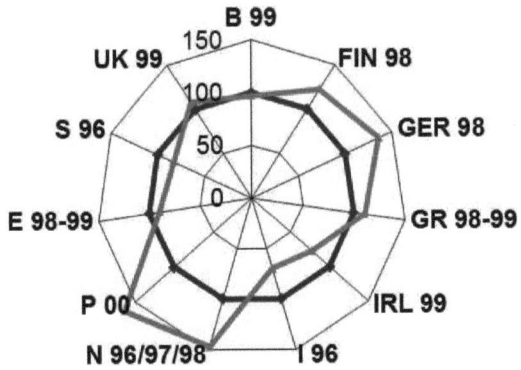

Figure 94a. Potatoes, by "elderly-2 members"

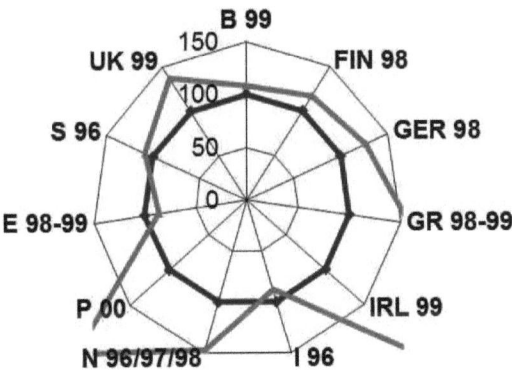

Figure 94b. Potatoes, by "adult-2 members"

ANNEX 3

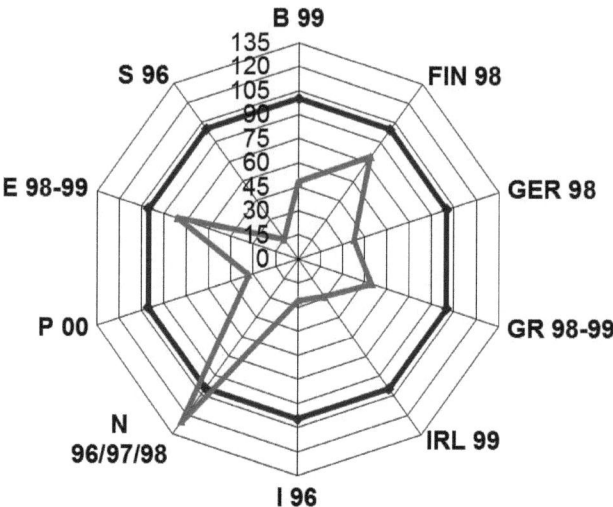

Figure 95a. Flour, by manual household head

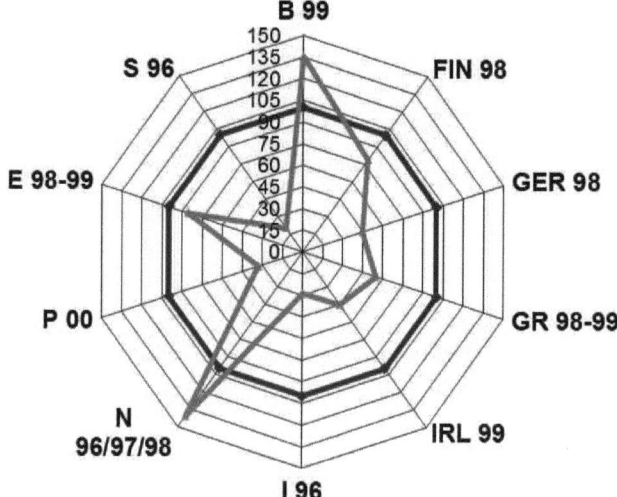

Figure 95b. Flour, by non manual household head

ANNEX 3

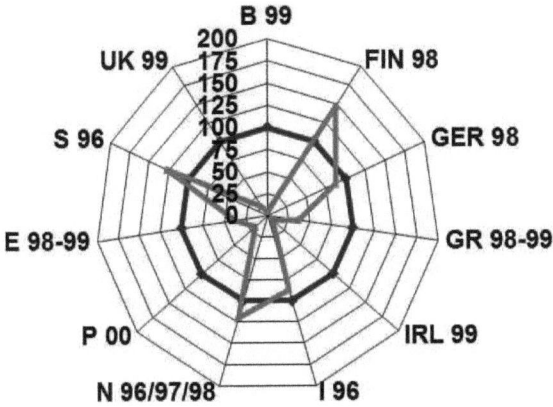

Figure 96a. Coffee in Rural areas

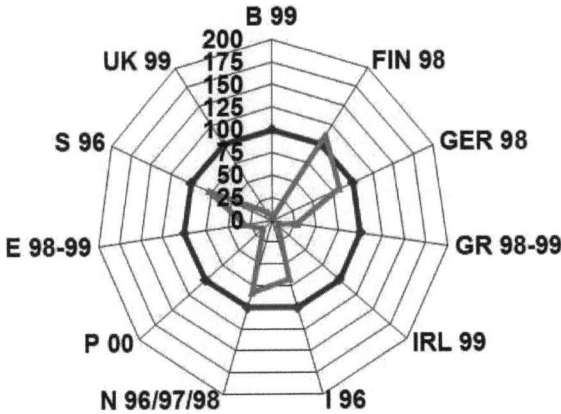

Figure 96b. Coffee in Urban areas

ANNEX 3

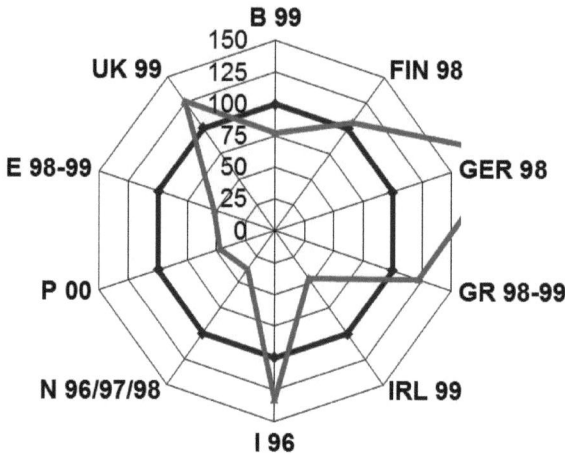

Figure 97a. Spirit in Rural areas

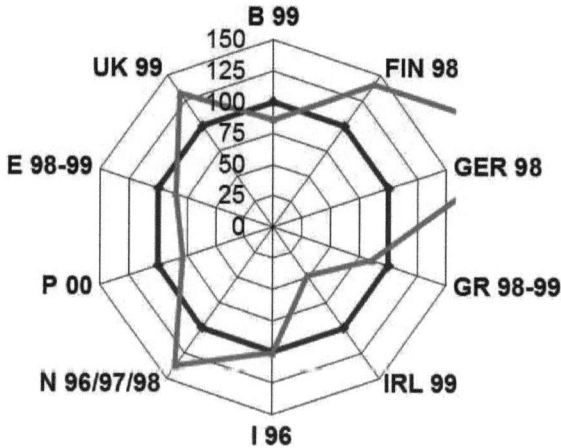

Figure 97b. Spirit in Urban areas

ANNEX 3

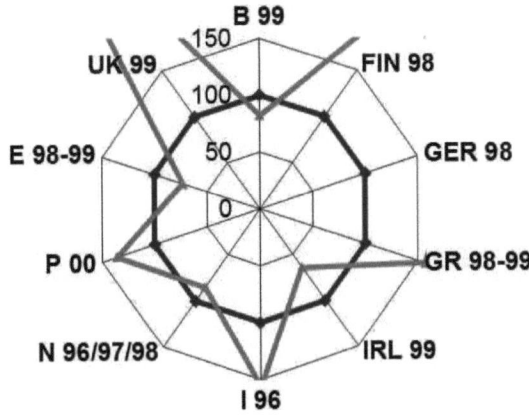

Figure 98a. Spirit, by "adult-single"

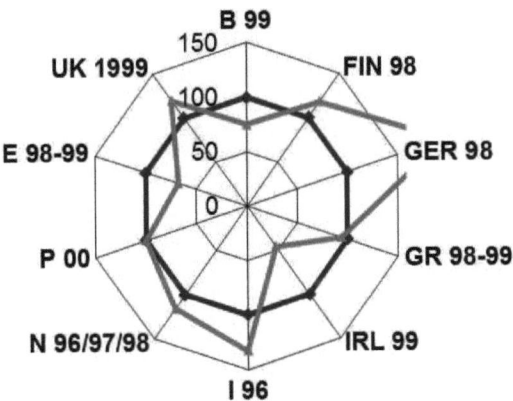

Figure 98b. Spirit, by "adult-2 members"

ANNEX 3

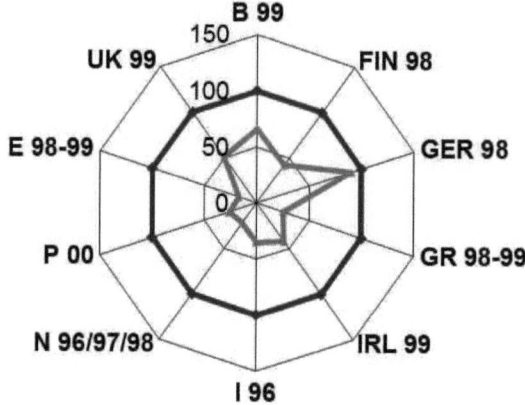

Figure 99a. Spirit, by "elderly-single" households

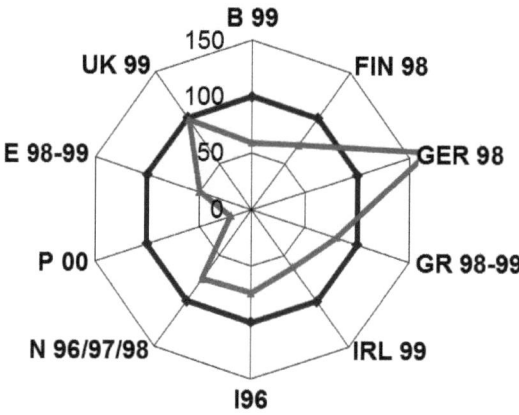

Figure 99b. Spirit, by "elderly-2 members" households

ANNEX 3

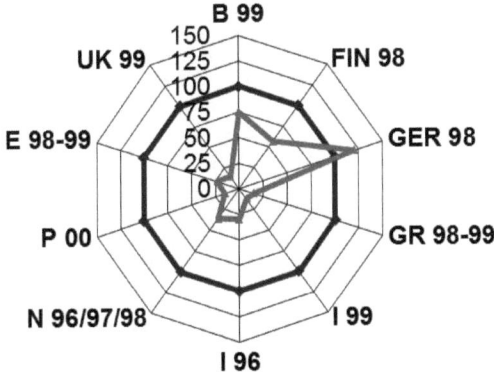

Figure 100a. Beer in Rural areas

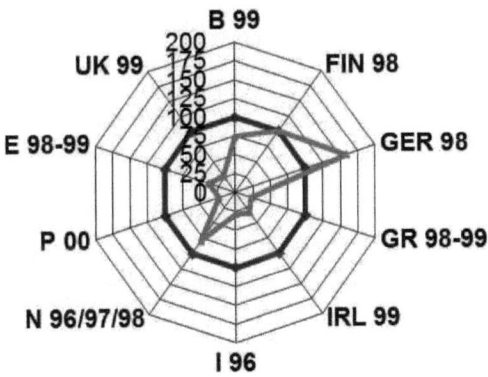

Figure 100b. Beer in Urban areas

ANNEX 3

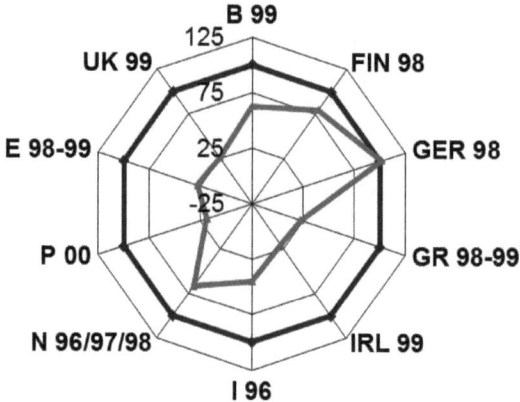

Figure 101a. Beer, by "adult-single" households

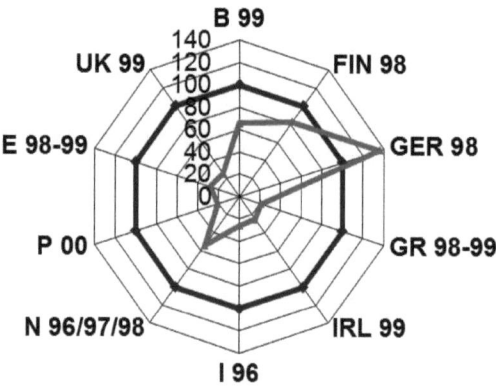

Figure 101b. Beer, by "adult-2 members" households

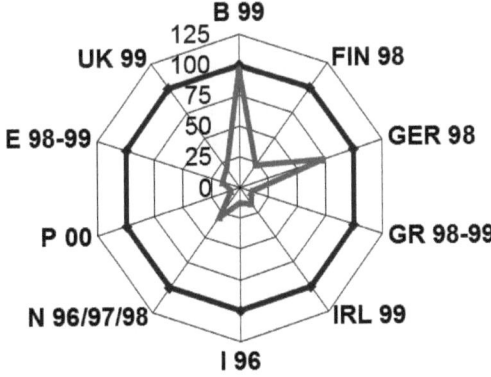

Figure 102a. Beer, by "elderly-single" households

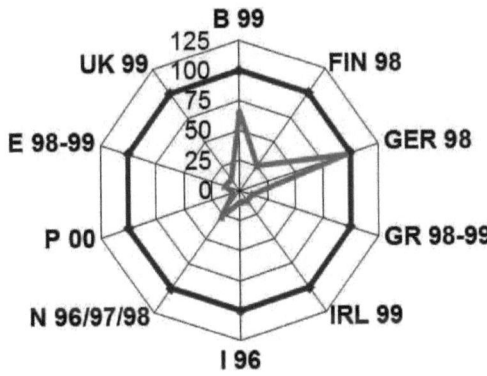

Figure 102b. Beer, by "elderly-2 members" households

ANNEX 3

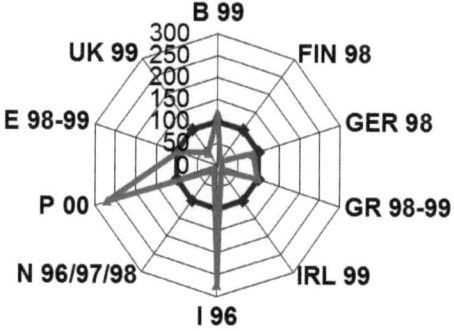

Figure 103a. Wine in Rural areas

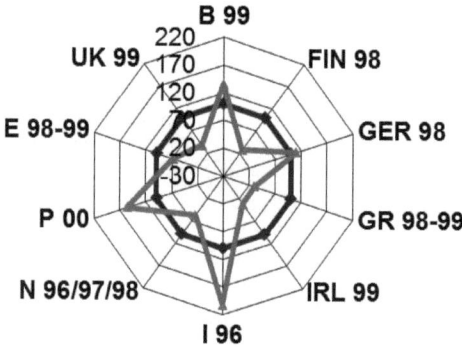

Figure 103b. Wine in Urban areas

ANNEX 3

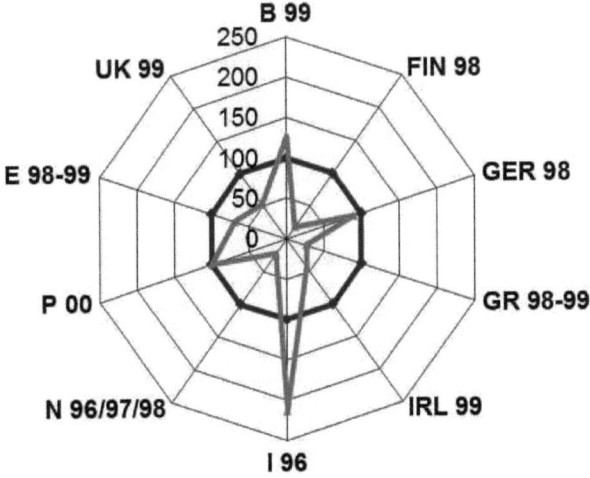

Figure 104a. Wine, by "adult-single" households

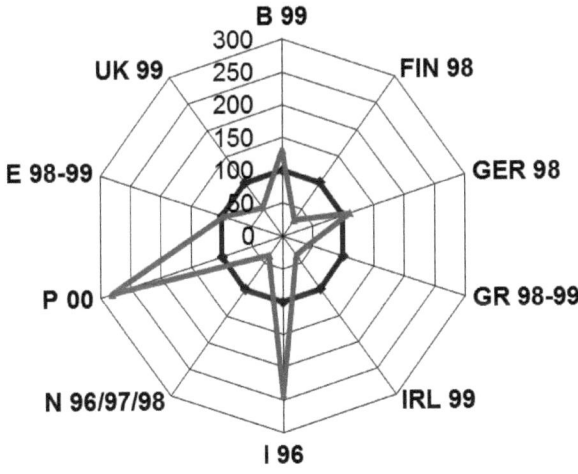

Figure 104b. Wine, by "adult-2 members" households

ANNEX 3

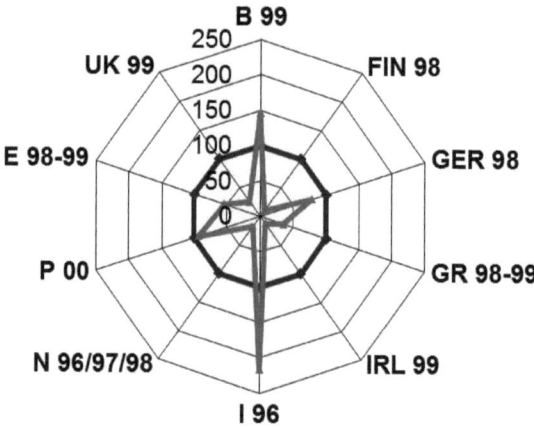

Figure 105a. Wine, by "elderly-single" households

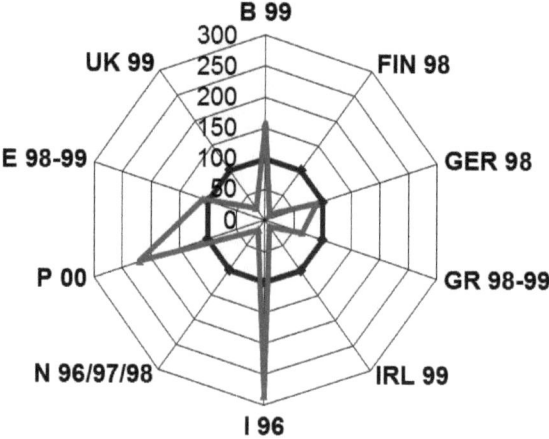

Figure 105b. Wine, by "elderly-2 members" households

ANNEX 3

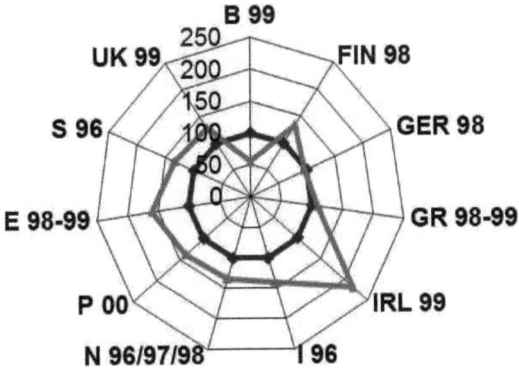

Figure 106a. Milk, by "adult-single" households

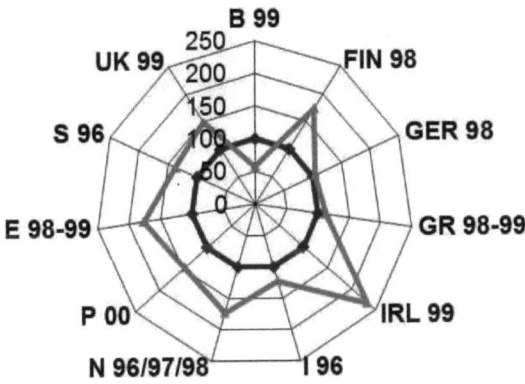

Figure 106b. Milk, by "adult-2 members" households

ANNEX 3

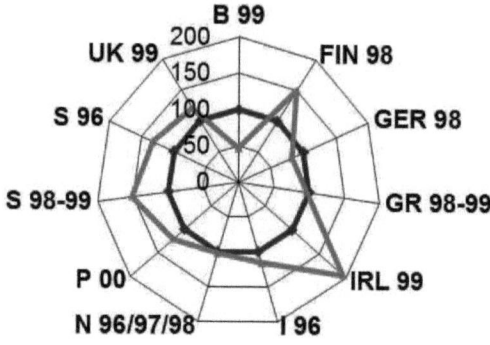

Figure 107a. Milk, by "elderly-single" households

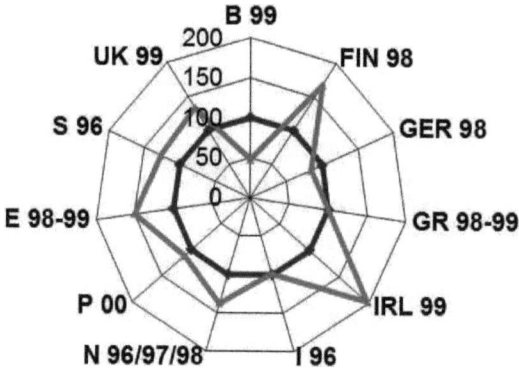

Figure 107b. Milk, by "elderly-2 members" households

ANNEX 3

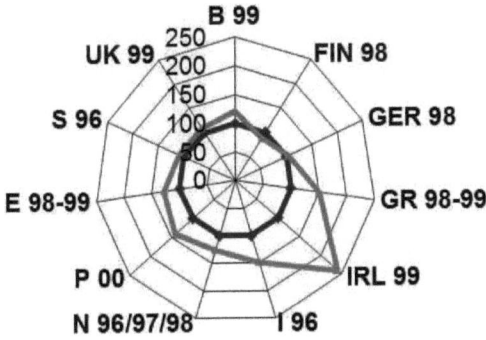

Figure 108a. Bread and Rolls, by "adult-single" households

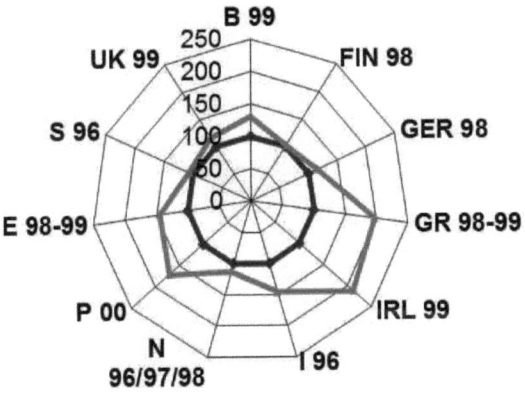

Figure 108b. Bread and Rolls, by "adult-2 members" households

ANNEX 3

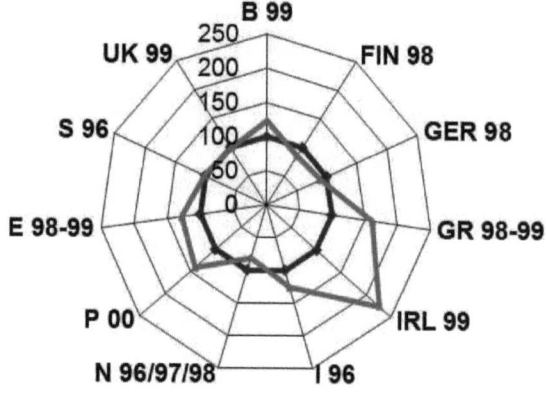

Figure 109a. Bread and Rolls, by "elderly-single" households

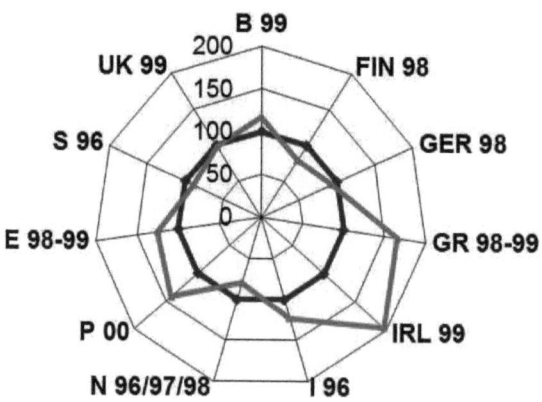

Figure 109b. Bread and Rolls, by "elderly-2 members" households

ANNEX 3

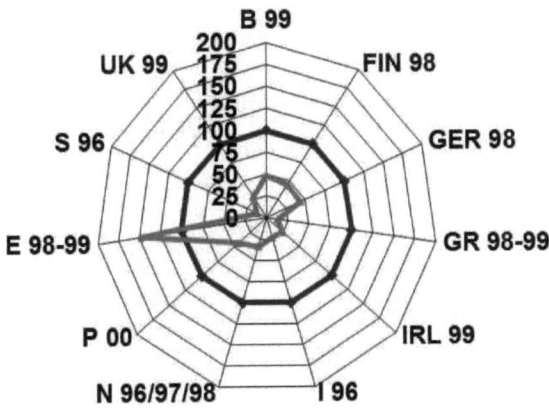

Figure 110a. Cocoa in Rural areas

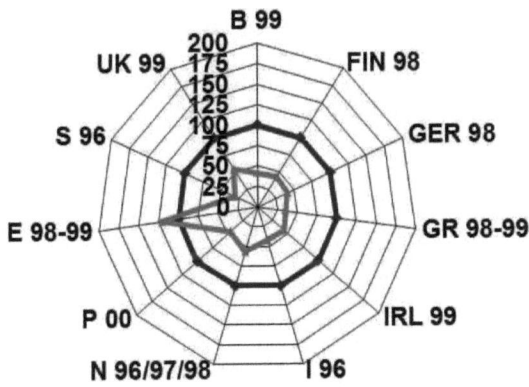

Figure 110b. Cocoa in Urban areas

ANNEX 3

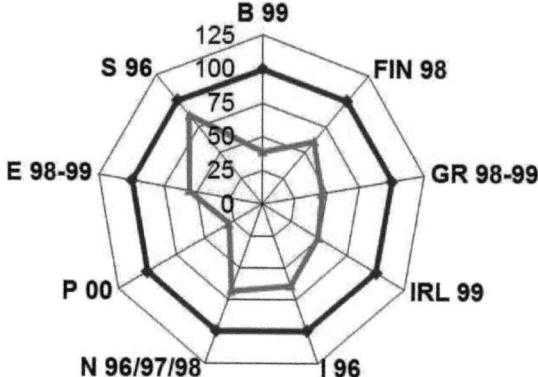

Figure 111a. Eggs, by Elementary education

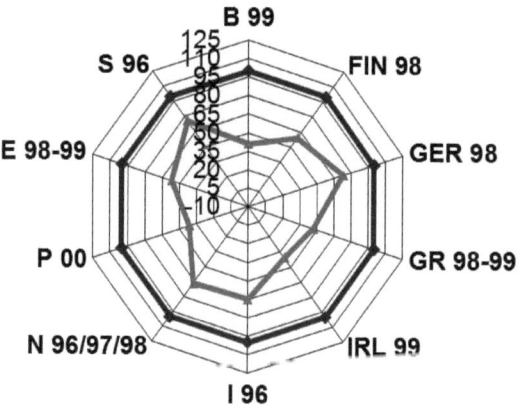

Figure 111b. Eggs, by Secondary education

ANNEX 3

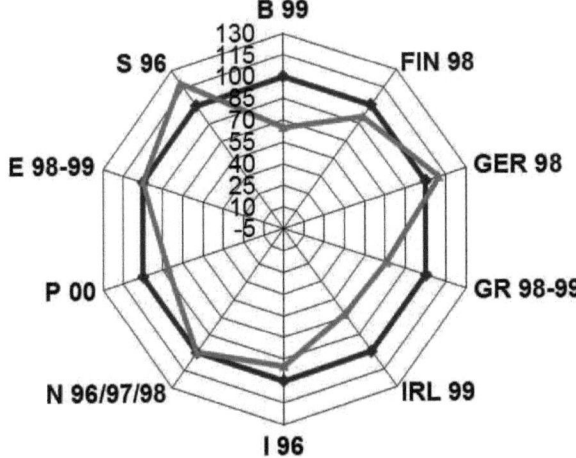

Figure 111c. Eggs, by Higher education

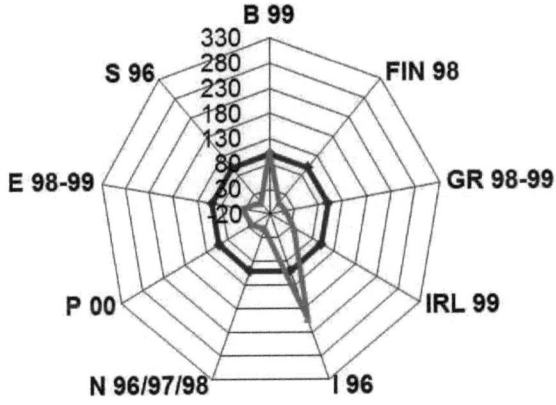

Figure 112a. Mineral water, by Elementary education

ANNEX 3

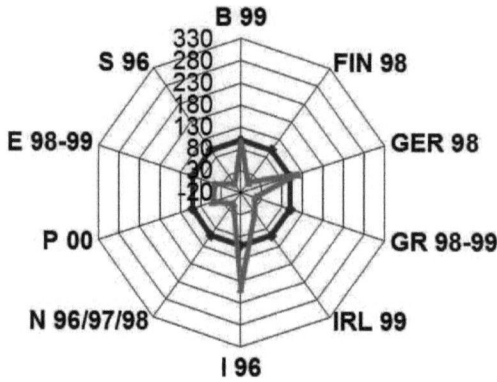

Figure 112b. Mineral water, by Secondary education

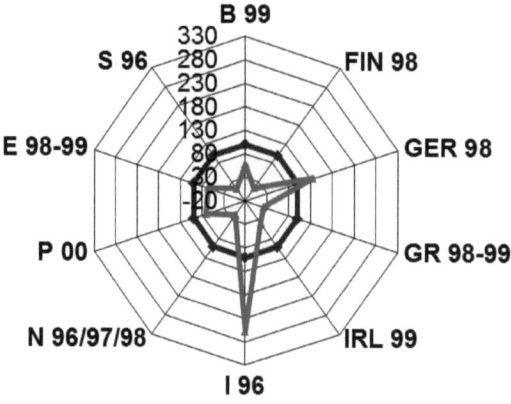

Figure 112c. Mineral water, by Higher education

ANNEX 4

Figure 113.a. Available information of occupation in the Austrian HBS

Code	Description of occupation	Skill level	Primary variable: Job description Code=P015	Secondary variable: Position in occupation Code=P014
I	II	III	IV	V
Details se below:				
Retired persons, unemployed persons etc. were classified among the occupation, which were finally practised				

I: Code
= 1 – 10

II: Description of occupation
a = Legislators, senior officials and managers, and professional
b = Technicians and associated professionals
c = Clerks, service workers, and shop and market sales workers
d = Skilled agricultural and fishery workers
e = Craft and related trades workers, and plant and machine operators and assemblers
f = Elementary occupations
g = Unemployed but economically active
h = Economically inactive (students, housewives, early retirements, etc.)
i = Pensioners
j = Armed Forces and others not elsewhere classified

Figure 113.a. Available information of occupation in the Austrian HBS
Code I, Code II

ANNEX 4

II: Description of occupation	III: Skill level	IV: Primary variable: Job description Code=P015
a	1	Codes= 02,03,04,05,06, 07,08,11
b	1	Codes= 09,10,12
c	2	Codes = 15,16,25
d	3 or 5	Codes=17,26
e	3 and 4	Codes= 13,14,20,21,22, 23,24
f	5	Codes= 18,19,27
g	(8)	
h	(8)	Codes= 28
i	8	
j	0 and 9	Codes= 00,01

Figure 113.b. Available information of occupation in the Austrian HBS Code II, Code III, Code IV

V: Secondary variable: Position in occupation Code=P014
0 – No details
1 – Farmer (smallholding)
2 – Farmer (farm of medium size)
3 – Farmer (farm of large size)
4 – Assistant at a farm (smallholding)
5 – Assistant at a farm (farm of medium size)
6 – Assistant at a farm (large size)
7 – Non – agricultural self –employer with an enterprise without employees
8 – Non-agricultural self- employer with an enterprise (1-4 employees)
9 – Non-agricultural self- employer with an enterprise (5-9 employees)
10 – Non-agricultural self-employer with an enterprise (10 or more employees)
11 – Assistant at a non-agricultural self-employed enterprise without employees
12 – Assistant at a non-agricultural self-employed enterprise (1-4 employees)
13 – Assistant at a agricultural self-employed enterprise (5-9 employees)
14 - Assistant at a agricultural self-employed enterprise (10 or more employees)
15 – Self employed person without college/university education

Figure 113.c. Available information of occupation in the Austrian HBS Code V, Position in Occupation 0 – 15

ANNEX 4

V: Secondary variable: Position in occupation **Code=P014**
16 – Self-employed person with college/university education
17 – Assistant at a self-employed person's enterprise without college/university education
18 – Assistant at a self employed person's enterprise with college/university education
19 – Apprentice in a worker's job
20 – Apprentice in a employee's job
21 – Unskilled workers (except agricultural and forestry)
22 – Unskilled workers in agricultural and forestry
23 – Semi-skilled workers
24 – Skilled workers
25 – Foreman/forewoman or master
26 – Employee - unskilled occupation
27 – Employee - skilled occupation
28 – Employee – medium occupation
29 – Employee - higher occupation
30 – Employee - high qualified occupation
31 – Employee - leading occupation
32 – Official -unskilled or semi-skilled occupation
33 – Official - skilled work or foreman occupation
34 – Official - unskilled work
35 – Official - simple occupation
36 – Official - medium occupation
37 – Official –high occupation
38 – Official – highly qualified
39 – Official – leading occupation
40 – Public employee of a contract –unskilled or semi-skilled occupation of a worker
41 – Public employee of a contract – skilled work or foreman occupation
42 – Public employee of a contract –unskilled occupation
43 – Public employee of a contract –simple occupation
44 – Public employee of a contract –medium occupation
45 – Public employee of a contract – high occupation
46 – Public employee of a contract – highly qualified or leading occupation
47 – never been gainful employment

Figure 113.d. Available information of occupation in the Austrian HBS Code V, Position in Occupation 0 – 15

P015	Job description

0 – no details or not classifiable profession
1 – Soldiers
2 – Members of legislative bodies and executive administration employees
3 – Managers and managing director in big enterprises
4 – Managers in small enterprises
5 – Physicists, mathematicians, engineer scientists
6 – Biological scientists and medics
7 – Scientific teachers
8 – Other scientists and related professions
9 – Technical qualified employees
10 – Qualified employees of biological science and health
11 – Non-scientific teachers
12 – Other qualified employees
13 – Office workers without customer service
14 – Office workers with customer service
15 – Professions in the services sector referring to persons and security men
16 – Models, salespersons and persons, who present something
17 – Qualified employees in agriculture and fishery
18 – Professions in winning minerals and building
19 – Metalworker, mechanics and related professions
20 – Precision workers, art craftsmen, printers and related professions
21 – Other skilled trades and related professions
22 – Operators stationary and related machines
23 – Operators of machines and fitters
24 – Drivers of a vehicle and operators of mobile machines
25 – Unskilled workers of services
26 – Unskilled agricultural fishery and related workers
27 – Unskilled workers of mining, trade of building, processing and transport
28 – Never been gainful employment

Figure 113.e. Available information of occupation in the Austrian HBS – Job description

ANNEX 5

ANNEX 5: CLASSIFICATION OF INDIVIDUAL CONSUMPTION BY PURPOSE (COICOP-HBS) [2]

00. TOTAL CONSUMPTION EXPENDITURE

01. FOOD AND NON-ALCOHOLIC BEVERAGES

01.1. Food

The food products classified here are those generally purchased for consumption at home. The group thus excludes food products normally sold for immediate consumption by hotels, restaurants, cafés, bars, kiosks, street vendors, automatic vending machines, etc. (11.1.1). Also excluded are cooked take-away dishes and the products of prepared-food suppliers and catering contractors even if they are delivered to the customer's home (11.1.1). Products sold specifically as pet foods are covered by (09.3.4).

01.1.1.   Bread and cereals (ND)

01.1.1.1.   Rice
- Rice in all forms, including rice prepared with meat, fish, seafood or vegetables. (d = kilo)

01.1.1.2.   Bread
- Bread and other bakery products such as crispbread, rusks, toasted bread, biscuits, gingerbread, wafers, waffles, crumpets, muffins and croissants. (d = kilo)

01.1.1.3.   Pasta products
- Pasta products in all forms, including pasta products containing meat, fish, seafood, cheese or vegetables. (d = kilo)

01.1.1.4.   Pastry-cook products
- Pastry-cook products such as cakes, tarts, pies, quiches, croquettes and pizzas.

01.1.1.5.   Sandwiches
- Any kind of bread (e.g. French loaf, pan loaf, croissants...) filled in by non-bread ingredients (e.g. ham, cheese, tuna, sausages, salad, vegetables...). (d = kilo)

---

[2] *Based on the final COICOP classification as prepared by the OECD after consultation with Eurostat, UNSD and the national statistical agencies of its Member Countries in October 1998 and approved in March 1999*

01.1.1.6. Other products

- Maize, wheat, barley, oats, rye and other cereals in the form of grains, flour or meal;
- other products such as malt, malt flour, malt extract, potato starch, tapioca, sago, other starches, cereal preparations (cornflakes, oat flakes, cereal bars, etc.) and dietary products and culinary ingredients based on flour, starch or malt extract.

Includes: couscous and similar farinaceous products prepared with meat, fish, seafood or vegetables; mixes and doughs for the preparation of bakery products or pastry-cook products.
Excludes: meat pies (01.1.2); fish pies (01.1.3); sweetcorn (01.1.7)

01.1.2. Meat (ND)

01.1.2.1. Fresh, chilled or frozen meat of bovine animals (d = kilo)

01.1.2.2. Fresh, chilled or frozen meat of swine (d = kilo)

01.1.2.3. Fresh, chilled or frozen meat of sheep and goat (d = kilo)

01.1.2.4. Fresh, chilled or frozen meat of poultry
- Chicken, duck, goose, turkey, guinea fowl, etc. (d = kilo)

01.1.2.5. Dried, salted or smoked meat and edible meat offal
- Sausages, salami, bacon, ham, pâté, etc. (d = kilo)

01.1.2.6. Other preserved or processed meat and meat preparations
- Canned meat, meat extracts, meat juices, meat pies, etc.
- mixture of meat of several types (beef, pork, sheep, ...) (d = kilo)

01.1.2.7. Other fresh, chilled or frozen edible meat
- Hare, rabbit and game (antelope, deer, boar, pheasant, grouse, pigeon, quail, etc.);
- horse, mule, donkey, camel and the like. (d = kilo)

Includes: meat and edible offal of marine mammals (seals, walruses, whales, etc.) and exotic animals (kangaroo, ostrich, alligator, etc.), animals and poultry purchased live for consumption as food.
Excludes: farinaceous products containing meat (01.1.1); frogs, land and sea snails (01.1.3); soups containing meat (01.1.9); lard and other edible animal fats (01.1.5).

01.1.3. Fish (ND)

01.1.3.1. Fresh, chilled or frozen fish (d = kilo)

ANNEX 5

01.1.3.2. Fresh, chilled or frozen seafood
- Crustaceans including land crabs, molluscs and other shellfish, land and sea snails, frogs
  (d = kilo)

01.1.3.3. Dried, smoked or salted fish and seafood (d = kilo)

01.1.3.4. Other preserved or processed fish and seafood and fish and seafood preparations
- Canned fish and seafood, caviar and other hard roes, fish pies, etc.
- mixture of fish and seafood of several types (d = kilo)

Includes: fish and seafood purchased live for consumption as food
Excludes: farinaceous products containing fish (01.1.1); fish soups (01.1.9)

01.1.4. Milk, cheese and eggs (ND)

01.1.4.1. Whole milk
- Raw, pasteurised or sterilised. (d = litre)

01.1.4.2. Low fat milk
- Raw, pasteurised or sterilised. (d = litre)

01.1.4.3. Preserved milk
- Condensed, evaporated or powdered (d = kilo)

01.1.4.4. Yoghurt (d = kilo)

Includes: yoghurt containing sugar, cocoa, fruit or flavourings

01.1.4.5. Cheese and curd (d = kilo)

01.1.4.6. Other milk products
- Cream, milk-based desserts, milk-based beverages and other similar milk-based products. (d = kilo)

Includes: milk and cream containing sugar, cocoa, fruit or flavourings, dairy products not based on milk such as soya milk *Excludes:* yoghurt containing sugar, cocoa, fruit or flavourings (01.1.4.4); butter and butter products (01.1.5)

01.1.4.7. Eggs
- Poultry eggs, egg powder and other egg products made wholly from eggs. (d = unit)

01.1.5. Oils and fats (ND)

ANNEX 5

01.1.5.1.	***Butter***
- Butter oil, ghee, etc. (d = kilo)	
01.1.5.2.	***Margarine and other vegetable fats***
- Including "diet" margarine and peanut butter. (d = kilo)	
01.1.5.3.	***Olive oil (d = litre)***
01.1.5.4.	***Edible oils***
- Corn oil, sunflower-seed oil, cottonseed oil, soybean oil, groundnut oil, walnut oil, etc. (d = litre)	
01.1.5.5.	***Other edible animal fats***
- Lard, etc. (d = kilo)	
Excludes: cod or halibut liver oil (06.1.1).	

01.1.6.	**Fruit (ND)**
01.1.6.1.	***Citrus fruits (fresh, chilled or frozen)***
- Orange, lemon, mandarin, grapefruit, etc. (d = kilo)	
01.1.6.2.	***Bananas (fresh, chilled or frozen) (d = kilo)***
01.1.6.3.	***Apples (fresh, chilled or frozen) (d = kilo)***
01.1.6.4.	***Pears (fresh, chilled or frozen) (d = kilo)***
01.1.6.5.	***Stone fruits (fresh, chilled or frozen)***
- Apricot, plum, peach, avocado, cherry, etc. (d = kilo)	
01.1.6.6.	***Berries (fresh, chilled or frozen)***
- Grapes, strawberries, etc. (d = kilo)	
01.1.6.7.	***Other fresh, chilled or frozen fruits***
- Other tropical fruits, melon, water melon, etc. (d = kilo)	
01.1.6.8.	***Dried fruit***
- Including fruit peel, fruit kernels, nuts and edible seeds. (d = kilo)	
01.1.6.9.	***Preserved fruit and fruit-based products***
- Dietary preparations and culinary ingredients based exclusively on fruit. (d = kilo)	
Excludes: vegetables grown for their fruit such as tomatoes, cucumbers and aubergines (01.1.7); jams, marmalades, compotes, jellies, fruit purées and pastes (01.1.8); parts of plants preserved in sugar (01.1.8); fruit juices (01.2.2); fruit concentrates and syrups for culinary use (01.1.9) or for the preparation of	

beverages (01.2.2).

01.1.7. Vegetables (ND)

01.1.7.1. Leaf and stem vegetables (fresh, chilled or frozen)
- Lettuce, chicory, endive, celery, cress, spinach, parsley, fennel, etc.

Excludes: asparagus (01.1.7.4)

01.1.7.2. Cabbages (fresh, chilled or frozen)
- Brocolis, cauliflower, etc. (d = kilo)

01.1.7.3. Vegetables cultivated for their fruit (fresh, chilled or frozen)
- Cucumber, tomato, aubergine, courgette, sweetcorn, beans, green pepper, pumpkins, olives etc. (d = kilo)

01.1.7.4. Root crops, non-starchy bulbs and mushrooms (fresh, chilled or frozen)
- Carrot, beet root, radish, turnip, onion, garlic, parsnip, leek, asparagus, artichoke, etc. (d = kilo)

01.1.7.5. Dried vegetables (d = kilo)

Includes: pulses

01.1.7.6. Other preserved or processed vegetables
- Vegetable-based products, dietary preparations and culinary ingredients based exclusively on vegetables.
- mixture of vegetables of several types (d = kilo)

Includes: gherkin with vinegar

01.1.7.7. Potatoes (d = kilo)

01.1.7.8. Other tubers and products of tuber vegetables
- Manioc, arrowroot, cassava, sweet potatoes and other starchy roots;
- Flours, meals, flakes, purees, chips and crisps, including frozen preparations such as chipped potatoes. (d = kilo)

Includes: sea fennel and other edible seaweed, other edible fungi (01.1.7.4)
Excludes: potato starch, tapioca, sago and other starches (01.1.1); soups, broths and stocks (01.1.9); ginger, pimento and other spices and condiments, culinary herbs (parsley, rosemary, thyme, etc.) (01.1.9); vegetable juices (01.2.2)

01.1.8. Sugar, jam, honey, chocolate and confectionery (ND)

01.1.8.1. Sugar
- Cane or beet sugar, unrefined or refined, powdered, crystallised or in lumps. (d

= kilo)

01.1.8.2. Jams, marmalades
- Including compotes, jellies, fruit purees and pastes, natural and artificial honey. (d = kilo)

01.1.8.3. Chocolate
- In bars or slabs. (d = kilo)

01.1.8.4. Confectionery products
- Chewing gum, sweets, toffees, pastilles and other.

01.1.8.5. Edible ices and ice cream
- Including sorbet. (d = litre)

01.1.8.6. Other sugar products
- Syrups and molasses, including parts of plants preserved in sugar;
- cocoa-based foods and cocoa-based dessert preparations;

Includes: artificial sugar substitutes (01.1.8.1)
Excludes: cocoa and chocolate-based powder (01.2.1); syrups for the preparation of beverages (01.2.2).

01.1.9. Food products n.e.c. (ND)

01.1.9.1. Sauces, condiments
- Seasonings (mustard, mayonnaise, ketchup, soy sauce, etc.), vinegar.

01.1.9.2. Salt, spices and culinary herbs
- Salt, spices, ginger, pimento and culinary herbs (parsley, rosemary, thyme etc.).

01.1.9.3. Baby food, dietary preparations, baker's yeast and other food preparations
- Homogenised babyfood and dietary preparations irrespective of the composition
- prepared baking powders, baker's yeast, dessert preparations such as vanilla aroma, pudding powders or dessert sauces, soup, broths, stocks, etc.

01.1.9.4. Other food products n.e.c.

Includes: soya substitutes of other food products except oil and soya milk
Excludes: milk-based dessert preparations (01.1.4); soya milk (01.1.4); artificial sugar substitute and cocoa-based dessert preparations (01.1.8).

01.2. Non-alcoholic beverages

ANNEX 5

> The non-alcoholic beverages classified here are those generally purchased for consumption at home. The group thus excludes non-alcoholic beverages normally sold for immediate consumption by hotels, restaurants, cafés, bars, kiosks, street vendors, automatic vending machines, etc. (11.1.1).

01.2.1.	Coffee, tea and cocoa (ND)
01.2.1.1.	***Coffee*** - Whether or not decaffeinated, roasted or ground, including instant coffee, coffee extracts and essences and coffee substitutes. (d = kilo)
01.2.1.2.	***Tea*** - Including maté and other plant products for infusions. (d = kilo)
01.2.1.3.	***Cocoa and powdered chocolate*** - Whether or not sweetened. (d = kilo)
Includes: preparations for beverages containing cocoa, milk, malt, etc.; coffee and tea substitutes; extracts and essences of coffee and tea.	

01.2.2.	Mineral waters, soft drinks, fruit and vegetable juices (ND)
01.2.2.1.	***Mineral or spring waters (d = litre)***
01.2.2.2.	***Soft drinks*** - Such as sodas, lemonades and colas (d = litre)
01.2.2.3.	***Fruit juices*** - Including syrups and concentrates for the preparation of beverages (d = litre)
01.2.2.4.	***Vegetable juices (d = litre)***
Includes: all drinking water sold in containers *Excludes*: non-alcoholic spirits, liqueurs, etc. (02.1.1); non-alcoholic wine, cider, etc. (02.1.2) and non-alcoholic beer (02.1.3).	

02.	**ALCOHOLIC BEVERAGES, TOBACCO AND NARCOTICS**

02.1.	Alcoholic beverages
The alcoholic beverages classified here are those generally purchased for consumption at home. The group thus excludes alcoholic beverages normally sold for immediate consumption by hotels, restaurants, cafés, bars, kiosks, street vendors, automatic vending machines, etc. (11.1.1).The beverages classified here include low or non-alcoholic varieties of beverages, which are generally alcoholic, such as non-alcoholic beer.	

ANNEX 5

02.1.1.	Spirits (ND)
02.1.1.1.	*Spirits and liqueurs.* - Mead; aperitifs other than wine-based aperitifs; non-alcoholic spirits, liqueurs, eaux-de-vie, etc. (d = litre)

02.1.2.	Wine (ND)
Wine from grapes or other fruit - Including cider and perry, fortified wine. (d = litre)	
02.1.2.1.	*Other* -Wine-based aperitifs, champagne and other sparkling wines, sake and the like. (d = litre)
Includes: non-alcoholic wine, etc.	

02.1.3.	Beer (ND)
02.1.3.1.	*Beer*
- All kinds of beer such as ale, lager and porter; - including low-alcoholic beer and non-alcoholic beer, shandy. (d = litre)	

Figure 114. COICOP – HBS 03 (Statistik Austria, 2003b)

I want morebooks!

Buy your books fast and straightforward online - at one of the world's fastest growing online book stores! Environmentally sound due to Print-on-Demand technologies.

Buy your books online at
www.get-morebooks.com

Kaufen Sie Ihre Bücher schnell und unkompliziert online – auf einer der am schnellsten wachsenden Buchhandelsplattformen weltweit!
Dank Print-On-Demand umwelt- und ressourcenschonend produziert.

Bücher schneller online kaufen
www.morebooks.de

OmniScriptum Marketing DEU GmbH
Heinrich-Böcking-Str. 6-8
D - 66121 Saarbrücken
Telefax: +49 681 93 81 567-9

info@omniscriptum.com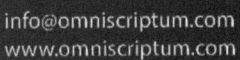
www.omniscriptum.com

Printed by Books on Demand GmbH, Norderstedt / Germany